MODELLING GROUNDWATER SYSTEMS

UNDERSTANDING AND IMPROVING GROUNDWATER

QUANTITY AND QUALITY MANAGEMENT

MODELLING GROUNDWATER SYSTEMS

UNDERSTANDING AND IMPROVING GROUNDWATER
QUANTITY AND QUALITY MANAGEMENT

DISSERTATION

Submitted in fulfilment of the requirements of
the Board for Doctorates of Delft University of Technology and of
the Academic Board of the UNESCO-IHE Institute for Water Education
for the Degree of DOCTOR
to be defended in public on
Tuesday, December 3, 2013 at 10:00 hours
in Delft, the Netherlands

by

Girma Yimer EBRAHIM

Master of Science in Hydroinformatics (with Distinction)
UNESCO-IHE, the Netherlands

born in Wolkite, Ethiopia

This dissertation has been approved by the supervisor
Prof. dr. ir. A. E. Mynett

Co-supervisor: Dr. A. Jonoski

Composition of Doctoral Committee:

Chairman	Rector Magnificus, Delft University of Technology
Vice-chairman	Rector, UNESCO-IHE
Prof. dr. ir. A. E Mynett	UNESCO-IHE/Delft University of Technology, supervisor
Dr. A. Jonoski	UNESCO-IHE, co-supervisor
Prof. dr. ir. A. B. K. van Griensven	UNESCO-IHE/Vrije Universiteit Brussel, Belgium
Prof. dr. ir. H. H. G. Savenije	Delft University of Technology
Prof. dr. ing. R. Hinkelmann	Technical University of Berlin, Germany
Prof. dr. ir. P. Seuntjens	Ghent University/ University of Antwerp, Belgium
Prof. dr. ir. N.C. van de Giesen	Delft University of Technology (reserve member)

CRC Press/Balkema is an imprint of the Taylor & Francis Group, and informa business

Published by:
CRC Press/Balkema
PO Box 11320, 2301 EH Leiden, the Netherlands
ISBN: 978-1-138-02404-5

Acknowledgments

This research would not have been possible without the support from a number of people. Everyone contributed in his/her own way, but I am most grateful to all.

First of all, I would like to express my deepest gratitude to my Promoter Prof. Arthur Mynett for his strong motivation, understanding, and wise guidance throughout the research and during the writing of the thesis. His positive and stimulating attitude has been a source of inspiration for me. Without his effort I would never have come to this stage.

I would like to express my sincere thanks and appreciation to my supervisors, Dr. Ann van Griensven and Dr. Andreja Jonoski. I am very grateful to Dr. Ann van Griensven for creating this PhD opportunity, for her guidance, and for her support to my family. I am also most grateful to Dr. Andreja Jonoski for his constant support, for trusting my capability to do independent research, and for providing valuable and constructive suggestions to complete my thesis.

Financial support for this research was provided by the EU/FP7 AQUAREHAB Project (Grant Agreement Nr.226565) for which I am very grateful. Additional support was provided by Dr. Ioana Popescu to whom I would like to offer my deep heartfelt thanks. Special thanks go to my friend Dr. Solomon Dachewu for his enormous support during this research and for always being available and willing to help me throughout my studies. Many thanks also go to Dr. Arlex Sanchez for his friendship and support. I would also like to thank Dr. Jan Broaders and Dr. Juliette Dujardin for sharing their model, and Dr. Dejonghe Winne and Dr. Britta Schmalz for providing their data.

The help of Prof. Dimitri Solomatine in using GLOBE and Delphi programming is much appreciated. I would like also to thank all staff members of the hydroinformatics chair group for sharing their knowledge, experience and support. My special thanks go to Giuliano Di Baldassarre for sharing his experience in paper writing and to Dr. Gerald Corzo for his much valued support in parallel computing. I am grateful to the hydrology chair group: to Dr. Yangxiao Zhou for his useful discussion, advice, and willingness to help me throughout my study; to Dr. Jochen Wenninger for his help with the on electrical resistance field measurements, and to Dr. Shreedar Maskey for our useful discussion and for sharing his computer code. I would like to thank Dr. Ali Maktumi, Dr. Ahmed Mushataq and Dr. Osman Abdella for their support during my stay in Sultan Qaboos University, Oman.

Acknowledgments

I would like to thank members of the doctoral examination committee for their reviews and very valuable comments and suggestions.

UNESCO-IHE has always been home away from home. I have made so many friends since my MSc study, many more than can be mentioned here. I would like to acknowledge all PhD fellows and MSc students with whom I shared social life and experience. The discussion with my colleagues contributed a lot to the progress of this research. Special thanks go to Maurizio, Mica, Kun Yan, Carlos, Mario, Nagendra, Anwar, Blagoj, Shah, Isnaneni, Silas, Oscar, Khalid, Yos, Patricia, Fiona, Guo, Zahara, Abel, Yasir, Veronica, Juliette, Heddy, Gaberla, Mijal, Adrian, Mona, Aline, and Salifu for their warm friendship.

Special thanks go to Lin Hoang who was working on her PhD in parallel with me and with whom I shared many experiences including suffering and joy during the entire period of our PhD studies. The road was long, with many ups and downs along the way, but together we managed. You helped me more than you are probably aware – thank you so much for your generous understanding and help.

I would like to express my gratitude to all my former and present Ethiopian colleagues at UNESCO-IHE, particularly Ermias, Mesge, Yared Abyeneh, Tseganeh, Adeye, Yenesewu, Feke, Yared Ashenafi, Seleshi, Rahel, Aki, Mule, Eden, Meseret, Abonesh, and Eskinder for the good times we shared on several occasions. Special words of thank go to my best friends Sirak and Tizitaw for their support and encouragement during the course of this study. My special thanks extend to Tedo who designed the cover page of this thesis.

Kind help and support was also obtained from numerous staff members at UNESCO-IHE. I would like to express my sincere gratituted to Jolanda Boots, Jos Bult, Ineke Melis, Sylvia van Opdorp - Stijlen, Maria Laura Sorrentino, Mariëlle van Erven, Anique Karsten, Peter Stroo and Peter Heerings. My strong appreciation goes to the UNESCO-IHE football practice and tournament organizers Davide, Klaas, Stefan, and Hermen.

Finally, I am very thankful to all my family members, relatives and friends in Ethiopia, in particular to my mother and father for their love, support and prayers, and to my brothers and sisters for their support and encouragement. My special thanks go to my brother Assfaw Yimer who is my source of confidence and who is the reason for who I am now. I am also very grateful to my wife's father and Azeb Assefa for the support they gave to my family during the period when I was away.

Above all, my special and warm thanks go to my wife Selamawit Assefa, for her love, patience and unlimited support during all this time. Special thanks and love to our son Dagim, and our daughter Dana who were born during this PhD study period. Dagi I am very well aware that I have to make up with you for my absence – and I promise I will. Dana I am much looking forward to seeing you in real life for the first time soon. Selam you endured multiple hardships raising Dagi and Dana alone. Without your encouragement, support and sacrifice my study would not have been possible. I only hope that someday I could reciprocate all the love you provided to me during hard moments.

Last but not least I praise the Almighty GOD for his love and blessings along all my ways.

This thesis is dedicated to my wife Selam, our son Dagim and daughter Dana.

Girma Yimer Ebrahim
Delft, the Netherlands

Summary

Groundwater is one of the most important natural resources. It is the principal source of drinking water in rural and many urban cities, and widely used for irrigation in most arid and semi-arid countries. However, recently it has become apparent that many human activities are negatively impacting both the quantity and quality of groundwater resources. In many parts of the world, groundwater resources are under increasing threat due to contamination and depletion by excessive pumping.

Groundwater contamination is the degradation of natural water quality as a result of human activities. Potential sources of groundwater contamination are numerous. Many of the contaminants seeping underground are substances that are routinely used and discarded by modern societies such as solvents used in the industries and fertilizer and pesticides applied in agricultural fields. Groundwater is said to be contaminated when contaminant concentration levels restrict its potential use. Once groundwater is contaminated, it may be difficult and expensive to clean up. Groundwater contamination incidents from major industries are becoming more common and are often the subject of major and expensive investigations and cleanup activities. Billions of dollars / euros are being spent every year on groundwater pollution problems in the U.S.A and in Europe. Such high costs have made this problem number one priority among environmental issues.

On the other hand, in response to increasing withdrawals to meet the needs of an expanding population, groundwater levels are continuing to decline around the world. Over-exploitation of aquifers that are in contact with the sea or other water bodies of inferior quality may result in deterioration of the aquifer's water quality. In coastal areas, where many of the world's largest cities are located, the available volume of fresh groundwater is reduced by seawater intrusion, due to over pumping. Over pumping also causes reduction in streamflow, potential losses of groundwater dependent ecosystems and land subsidence.

Groundwater and surface water (GW-SW) are interconnected and interdependent in almost all hydrological settings. Much of the flow in streams and water in lakes and wetlands is sustained by the discharge of groundwater, while surface waters provide recharge to groundwater in other settings. Water fluxes between groundwater and streams mediate the transport of contaminants between these two hydrologic compartments. Understating GW-SW interaction has long been a topic of interest among hydrologists mainly due to its role in conjunctive use, contaminant transport, riparian zone management, etc. However, there are still many difficulties to obtain reliable estimates of the spatial and temporal distribution of fluxes.

One way to address these groundwater management challenges is through the use of models. Models allow the analysis of present conditions of the groundwater systems as well as its temporal development. From the water quality perspective models are useful tools to understand and predict the behaviour of contaminants in flowing groundwater. This helps to reliably assess the risks arising from groundwater contamination problems, and to design alternative remediation measures. Models are also useful tools to understand the dynamics of GW-SW fluxes exchange and associated biogeochemical processes over a wide range of spatial and temporal scales. Also, a more detailed analysis geared toward management and protection of groundwater and surface water resources can be achieved with models.

The overall objective of this thesis is to investigate and highlight the value of combining different modelling techniques, to asses and evaluate different aspects of complex groundwater resources problems. With different case studies, the unique applications of combined models as analysis and management tools for groundwater problems are investigated. The first part of this thesis deals with the assessment of natural attenuation as a remedy for chlorinated solvent contaminated groundwater. The study area is the Vilvoorde-Machelen site which is one of the oldest Belgian industrial centres characterized by regional aquifer contamination mainly due to chlorinated aliphatic hydrocarbons (CAHs). CAHs originating from multiple sources in this site are flowing towards the Zenne River and posing risk to the environment. Three dimensional groundwater flow model (MODFLOW) and three dimensional multi-species reactive transport model (RT3D) was used to investigate the migration of these contaminants at the site, and to evaluate the potential of natural attenuation at the site to remediate CAHs. Results of this study showed that, although natural attenuation is occurring at the site, the estimated biodegradation rates are not significant to reduce the concentration of CAHs to acceptable level in near future. The concentrations of CAHs can be reduced significantly when the natural attenuation was combined with the removal of the pollution source. However even this reduction is not enough to reduce the concentrations of the contaminants to the levels required by environmental standards. Hence, other remedial alternatives such as enhanced biodegradation should be investigated further. This case study demonstrated some possible analyses of these problems that can be carried out with such modelling approaches when limited data are available, especially about the pollution source.

The second part of this thesis investigated the local scale GW-SW interactions in the Zenne River. A number of studies have explored the effects of spatial and temporal variability of GW-SW flux exchanges in different hydrologic setting using temperature and water level measurements. However, the effect of temporal resolution of water level and temperature data on estimating fluxes, remains largely unresolved. Therefore, this study investigated the effect of temporal resolution of input data on temporal variation of GW-SW flux exchanges. Understanding and accurately quantifying GW-SW flux exchanges at local scale is mainly required to evaluate the attenuation capacity of streambed sediments where CAHs contaminated groundwater is flowing through. GW-SW fluxes exchange was simulated using a variably saturated two-dimensional groundwater flow and heat transport model (VS2DH) calibrated using hourly and daily water level and temperatures measured at multiple depths below the riverbed. Results of the study showed that while the hourly model characterizes all the dynamics of GW-SW interactions, the daily model fails to represent this dynamics, particularly during high river flow events. Therefore, when simulating GW-SW interaction at a local scale, it is important to give attention to the temporal resolution of data used for model calibration and validation.

The third part of the modelling effort of this thesis involves quantifying GW-SW interaction in riparian wetlands. The first step in developing nutrient transport model is to develop a flow model. In the riparian wetland, Kielstau, North Germany, there is growing interest to investigate the nutrient transport and transformation processes in the riparian wetland. Several studies and monitoring campaigns were carried out in the past, all aiming at hydrological and water quality conditions of the local area. However, the hydrologic connection between the riparian wetland and the river, as well as the existing drainage ditches, is still poorly understood. This hampers the understanding of the mobility and transformation of nutrients in the wetland. For this reasons a combined modelling approach was designed and implemented with a transient two-dimensional horizontal groundwater model (MODFLOW) and an analytical profile model (STWT1). The MODFLOW model was calibrated using weekly measured water levels, whereas the STWT1 model applied at one of the river-wetland cross-section is calibrated using daily water level measurements. Results showed that while the MODFLOW model is able to simulate the velocity fields and residence time in the riparian wetland, it does not quite capture the exchange dynamics at the river-wetland interface as observed using STWT1 model. Therefore, daily to sub-daily measurements are needed to understand the flow (and in subsequently nutrient) dynamics particularly at the river-wetland interface.

Summary

The fourth part of this thesis deals with optimal management of shallow groundwater levels in drained riparian wetland. The application case study again is Kielstau, North Germany. Excess water from the riparian wetland is drained off using open drainage ditches of different length and size. The water level in the drains needs to be maintained at appropriate level to assure the effectiveness of the riparian wetland in removing nutrients. The study investigates the applicability of a simulation-optimization approach for water level management to identify optimal drain elevations and water levels that allow traffic for small machinery as well as minimize drainage outflow from the shallow aquifer. Two global optimization algorithms known as genetic algorithm (GA), and controlled random search (CS4) were combined with the simulation groundwater model (MODFLOW) to find the optimal water levels in the existing drains. The results are in mutual agreement, which suggests that the water levels are near optimal solutions. In general it can be concluded that incorporating management goals and constraints significantly improves the MODFLOW modelling approach.

The final part of the study in this thesis involves assessing the feasibility of managed aquifer recharge (MAR). In arid countries like Oman, the problem of increasing seawater intrusion has raised significant interest in using MAR for augmenting groundwater supplies and for preventing further seawater intrusion. Since freshwater resources in Oman are scarce, treated municipal wastewater is considered as a main source to restore or enhance groundwater resources during periods of surplus for later use. To assess the feasibility of MAR in the Samail Lower Catchment, Oman a simulation-optimization approach was used. The objective is to determine maximum recharge and extraction rates while meeting system and operational constraints. By coupling the existing simulation model with a successive linear programming optimisation model and NSGA-II multi-objective optimisation algorithm, the optimal recharge and extraction pumping schemes (number and locations of wells, rates of injection and pumping) were determined. Results show that the aquifer is capable of absorbing large volumes of water and the recovery efficiency is high enough to consider MAR as a feasible option. By maintaining adequate freshwater storage, salt-water intrusion can be reduced or even inverted.

Overall, the thesis has demonstrated the capabilities and limitations of combining different modelling techniques to achieve optimal groundwater management. Focusing on one single modelling approach is hardly ever sufficient for achieving optimal solutions to groundwater management problems. New guidelines on using combined models are clearly needed and it is to these needs that this thesis has provided its main contribution.

Samenvatting

Grondwater is een van de belangrijkste natuurlijke grondstoffen. Het is de voornaamste bron van drinkwater in veel landelijke als stedelijke gebieden en is vaak de enige bron voor irrigatie in aride en semi-aride gebieden. Het wordt echter steeds duidelijker dat menselijke activiteiten een negatieve invloed hebben op zowel kwantiteit als kwaliteit van grondwaterbronnen. In veel gebieden in de wereld wordt het grondwater bedreigd door overexploitatie en verontreiniging.

Er zijn vele oorzaken voor grondwater verontreiniging: van huishoudelijke afvalstoffen die in de ondergrond wegsijpelen, tot oplosmiddelen gebruikt door de industrie of bestrijdingsmiddelen en meststoffen uit de landbouw. De mate van vervuiling wordt bepaald door de concentratiegraad van de vervuiling – als deze eenmaal de norm heeft overschreden, dan wordt het moeilijk en duur om de vervuiling ongedaan te maken. Grondwater vervuiling ten gevolge van ongelukken in belangrijke industrieën komen steeds vaker voor en vereisen dure tegenmaatregelen. Elk jaar worden in de USA en Europa miljarden besteed om grondwater vervuiling tegen te gaan. Vanwege de hoge kosten die hiermee gemoeid zijn heeft het voorkomen van grondwater verontreiniging dan ook de hoogste prioriteit bij het nemen van milieu maatregelen. Desalniettemin blijft de grondwaterstand dalen op veel plaatsen in de wereld, vanwege de toenemende vraag ten gevolge van een steeds groeiende wereldbevolking. Overexploitatie van grondwater reservoirs in kustgebieden leidt tot verzilting en afname van de grondwaterkwaliteit, met als gevolg verlies aan ecosystemen en – soms aanzienlijke – bodemdaling.

Grondwater en oppervlaktewater zijn nauw met elkaar verbonden beïnvloeden elkaar op hydrologisch gebied. Veel van het oppervlaktewater komt vanuit het grondwater, terwijl omgekeerd het oppervlaktewater de grondwater voorraden aanvult. De uitwisseling tussen grondwater en oppervlaktewater is direct van invloed op het transport van verontreiniging. Vandaar dat de onderlinge interactie door veel hydrologen wordt bestudeerd, met name ten behoeve van het beheer van oeverzones. Er zijn echter nog veel aspecten onbekend op het gebied van ruimtelijke en tijdsafhankelijke uitwisselingsprocessen.

Een manier om deze processen te bestuderen is met behulp van wiskundige modellen. Deze maken het mogelijk om de huidige situatie in kaart te brengen en beter te begrijpen welke ontwikkelingen in de toekomst mogelijk zijn. Zo kan de verspreiding van verontreiniging in het grondwater worden nagegaan teneinde risico's te kunnen inschatten en tegenmaatregelen te kunnen nemen. Modellen kunnen ook gebruikt worden om de uitwisselingsprocessen tussen grond- en oppervlaktewater beter te begrijpen inclusief de biochemische processen op verschillende tijd/ruimteschalen. Ook kan het effect van beheersmaatregelen en beschermingsmaatregelen beter worden nagegaan.

Het doel van dit proefschrift is om na te gaan of het mogelijk en zinvol is om verschillende modelleertechnieken te combineren met het doel om een beter inzicht te verkrijgen in complexe grondwaterproblemen. Aan de hand van verschillende case studies wordt de geschiktheid nagegaan van het gebruik van gecombineerde modellen voor analyse en beheer van grondwaterproblemen. In het eerste deel van dit proefschrift wordt nagegaan of natuurlijke afbraak een oplossing kan zijn voor het verwijderen van chemische oplosmiddelen uit het grondwater. Het studiegebied betreft een locatie nabij Vilvoorde-Machelen in België, een van de oudste Belgische industriële gebieden waar een lokale grondwater lens vervuild is door voornamelijk gechlorineerde aliphatische koolwaterstoffen (CAHs). Deze worden door verschillende bronnen aangevoerd en stromen in de richting van de rivier de Zenne waardoor ze een bedreiging vormen voor het milieu.

Het 3D grondwatermodel (MODFLOW) en het 3D reactieve transportmodel (RT3D) zijn gebruikt om de verspreiding van de verontreinigingen te onderzoeken en de mogelijkheden na te gaan van natuurlijke afbraak om de CAHs te verminderen. De resultaten lieten zien dat er weliswaar sprake is van enige natuurlijke afbraak in het betreffende gebied, maar dat deze verre van voldoende is om de CAHs in de nabije toekomst tot een acceptabel niveau te reduceren. De concentratie van CAHs kan wel aanzienlijk worden gereduceerd door natuurlijke afbraak in combinatie met het verwijderen van de bron. Maar zelfs in dat geval worden de milieu normen niet gehaald. Vandaar dat andere technieken zoals versnelde biodegradatie nader onderzoek verdienen. Deze casus liet zien dat het goed mogelijk is om dergelijke milieuproblemen te onderzoeken, zelfs als er slechts beperkte meetgegevens beschikbaar zijn van met name de bron van de verontreiniging.

In het tweede deel van dit proefschrift zijn de lokale interacties tussen grondwater en oppervlaktewater in de rivier de Zenne bestudeerd. Op basis van temperatuurmetingen in het veld zijn de uitwisselingsfluxen onderzocht en is de variabiliteit in ruimte en tijd in kaart gebracht. Het effect van tijdsafhankelijkheid van de fluxen blijkt echter moeilijk te onderzoeken. Daarom is gekeken naar het effect van tijdresolutie van de invoergegevens op de uitwisselingsfluxen. Daarbij bleek dat een betere beschrijving van de processen op lokale schaal nodig is om acceptabele resultaten te verkrijgen van het zelfreinigend vermogen van sediment in rivierbeddingen. In dit proefschrift is een 2D model voor grondwaterstroming en warmtetransport (VS2DH) gebruikt om de uitwisselingsfluxen numeriek te simuleren. Het model is gekalibreerd aan de hand van uurlijkse en dagelijkse temperatuurmetingen op meerdere dieptes in het rivierbed. De resultaten geven aan dat het model met tijdstappen van een uur de dynamische processen goed weergeeft, maar met tijdstappen van een dag niet meer, met name tijdens hoogwater afvoeren. Voor het juist simuleren van uitwisselingsprocessen op lokale schaal is het belangrijk om het model te kalibreren en valideren aan de hand van meetgegevens met een hoge temporele resolutie.

Het derde deel van dit proefschrift betreft het modelleren van grondwater-oppervlaktewater uitwisseling in oeverzones van wetlands. Allereerst is een stromingsmodel ontwikkeld voor het studiegebied nabij Kielstau in Noord Duitsland. Daar waren al verscheidene studies uitgevoerd naar de hydrologische condities en aspecten van water kwaliteit, maar nog niet naar de uitwisselingsfluxen tussen grond- en oppervlaktewater. Daartoe is in dit proefschrift een gecombineerde modelaanpak opgezet met enerzijds een tijdsafhankelijk twee-dimensionaal grondwatermodel (MODFLOW) en anderzijds een analytisch profielmodel (STWT1). Het MODFLOW model is gekalibreerd aan de hand van gemeten wekelijkse waterstanden, het STWT1 profielmodel met behulp van dagelijkse waarnemingen. De resultaten laten zien dat MODFLOW wel de snelheidsvelden en verblijfstijden kan simuleren, maar niet de uitwisseling zoals STWT1 dat kan. Om dat te bereiken zijn (twee maal) dagelijkse waarnemingen vereist, met name op het grensvlak tussen rivier en wetland.

Het vierde deel van dit proefschrift gaat over het optimaal beheer van het ondiepe grondwater in gedraineerde oeverzones. De casus hier is wederom het gebied bij Kielstau in Noord Duitsland waar kanalen en sloten worden gebruikt om overtollig water af te voeren. De waterstand in de sloten moet op het juiste niveau worden gehouden om te zorgen dat de oeverzones hun reinigende werking kunnen doen. In dit onderzoek is nagegaan of hier een gecombineerde optimalisatie-simulatie aanpak kan worden toegepast die de optimale ligging en waterstanden van de sloten regelen. Daartoe werden twee optimalisatietechnieken, Genetisch Algoritme (GA) en Controlled Random Search (CS4), gecombineerd met het grondwatermodel MODFLOW. De resultaten van beide aanpakken kwamen goed overeen, wat leidt tot de conclusie dat de waterstanden dicht bij de optimale oplossing lijken te zijn. Als algemene conclusie geldt dat het meenemen van beheers doelstellingen en bij het gebruik van MODFLOW tot aanzienlijk betere oplossingen leidt.

Het laatste deel van dit proefschrift betreft onderzoek naar de mogelijkheid van optimaal ondergronds water beheer. In droge landen zoals Oman met toenemende verzilting van zoetwater reservoirs in kustgebieden bestaat hieraan grote behoefte. Aangezien zoetwaterbronnen hier schaars zijn, wordt overwogen om gezuiverd afvalwater in de ondergrondse reservoirs te pompen in tijden van overschot, zodat die later weer kunnen worden teruggewonnen.

Om de haalbaarheid van een dergelijke aanpak na te gaan voor de Samail Lower Catchment in Oman, is wederom een simulatie-optimalisatie aanpak gehanteerd. Het doel was om vast te stellen wat de maximale pomp- en extractie capaciteit kan zijn. Daartoe is een bestaand grondwatermodel gekoppeld met een lineair programmeer model teneinde een optimaal pompschema vast te stellen (aantal en locatie van bronnen en putten, debieten van de pompen, etc.). De resultaten lieten zien dat de acquifer in staat was om grote hoeveelheden water te absorberen en dat het opnieuw onttrekken efficiënt genoeg is om optimalisatie van grondwaterbeheer in de praktijk toe te passen. Door steeds voldoende grondwater in het reservoir te houden wordt verdere de verzilting tegengegaan en zelfs terug gedrongen.

In het algemeen kan worden vastgesteld dat dit proefschrift heeft aangetoond wat de mogelijkheden en beperkingen zijn van het combineren van verschillende modelleertechnieken voor optimaal grondwater beheer. In de meeste gevallen is het gebruik van slechts een enkel model daarvoor niet genoeg. Nieuwe richtlijnen zijn nodig aangaande het gebruik van gecombineerde modellen. Dit proefschrift heeft getracht daaraan een bijdrage te leveren.

Contents

Contents

Contents

1 Introduction

1.1 Background

Groundwater is an important source of water supply. As a water supply source, groundwater has several advantages when compared to surface water: it is generally of higher quality, better protected from pollution, less subject to seasonal and perennial fluctuations, and more uniformly spread over large regions than surface water (Zektser and Lorne, 2004). There are many countries in the world where groundwater is one of the major sources of drinking water (e.g., Denmark, Malta, and Saudi Arabia). Groundwater also provides the largest amount of the total water resources in other countries. For example, in Tunisia it is about 95% of the total water resources, in Belgium it is 83%, in the Netherlands, Germany and Morocco it is 75% (Zektser and Lorne, 2004). Groundwater is also a major source of industrial and agricultural uses in many countries. Continuously increasing development has led to overexploitation of groundwater resources and growing impacts of human actives on aquifers in many countries, such as decline in groundwater levels and deterioration of groundwater quality. Groundwater is said to be contaminated when contaminant concentration levels restrict its potential use. Groundwater pollution caused by human activities can be broadly classified as point-source and non-point-source pollution. Point-source pollution refers to contamination originating from a single contamination site. Industrial waste disposal sites, accidental spills, leaking gasoline storage tanks, and dumps or landfills are examples of point sources. Some contamination problems originate from spatially distributed sources. Chemicals used in agriculture, such as fertilizers, pesticides and herbicides are examples of non-point-source pollution as they spread out across wide areas.

Once groundwater is contaminated, it may be difficult and expensive to clean up. This contamination not only threatens public health and the environment, but it costs the public large amounts of money for clean up (Granholm and Chester, 2003). Sources of groundwater contamination are numerous and are as diverse as human activities (Figure 1.1). Many sites, contaminated to various degrees by several chemicals or pollutants, exist in a number of geographical locations. The management of these contaminated sites is a great concern in view of the risks associated with such sites. While there are many causes of groundwater contaminations, the three that are most common are chlorinated solvents, agricultural chemicals and saltwater intrusions. This research focuses on evaluating best available approaches for protecting groundwater and surface water from these selected groundwater pollution problems using combined modelling approach. Currently, it is commonly accepted that dealing with different groundwater flow and pollution problems may require the use of combinations of modelling approaches, but insights and guidelines on the suitable approaches for specific problems are very much needed. Therefore, the following presentation will first introduce the characteristics of the three main problem areas addressed in this thesis, followed by the available modelling approaches, leading to the identification of the research objectives.

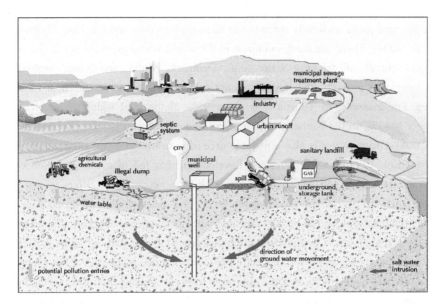

Figure 1.1: Sources of groundwater contaminations (Source: Zaporozec and Miller (2000)).

1.1.1   Chlorinated solvent contaminated groundwater

Groundwater contamination incidents from major industrial complexes are becoming more common and are often the subject of major, expensive investigations and clean-up activities. Groundwater pollution by chlorinated solvents (chlorinated aliphatic hydrocarbons (CAHs)) such as Tetrachloroethylene (PCE) also known as, perchloroethylene and trichloroethene (TCE) represents a serious environmental problem in many industrial areas (Pankow and Cherry, 1996). These compounds represent the most troublesome class of organic contaminants found in urban groundwater to-date. Concern about contamination with PCE or TCE has become more acute, owing to the observation that certain anaerobic bacteria in groundwater are able to transform these compounds to more hazardous chemicals such as vinyl chloride (VC) which is known to be carcinogenic in humans (EPA, 2000). PCE and TCE are only slightly soluble in water and thus may exist as nonaqueous phase liquids (NAPLs). Since they are denser than water they are referred as dense non-aqueous-phase liquid (DNAPL). When these chemicals spill and infiltrate through the vadose zone, a portion of it may be trapped and immobilized within the unsaturated porous formation in the form of blobs or ganglia (Mercer and Cohen, 1990). Given that the pressure head at the capillary fringe is sufficiently large, upon reaching the water table the DNAPL continue to sink leaving behind trapped ganglia until they encounter an impermeable layer, where a pool with relatively small cross-section starts to form. DNAPL remaining trapped in the soil-aquifer matrix and pools formed in the impermeable layer acts as continuing sources of dissolved contaminants to groundwater, and greatly complicate soil and aquifer remediation. Chlorinated solvents exhibit chemical and physical characteristics that enable them to contaminate vast quantity of groundwater. Physical and chemical properties of the most commonly encountered chlorinated solvents, which are also interest of this study, are summarized in Table 1.1. As the number of chlorine atoms decreases, molecular weight and density generally decreases, whereas vapor pressure and solubility is increasing. Highly chlorinated compounds such as, PCE and TCE exhibit moderate sorption, causing considerable retardation. The low solubility of PCE and TCE hinders the mixing of these compounds with the water in the contaminated aquifer. Low solubility and relatively high density are key properties that lead to the complexity of their distribution in the subsurface; consequently, billions of dollars are being spent in efforts to remediate groundwater contamination from chlorinated organic compounds (Chien et al., 2004). Hence, identification and removal of chlorinated solvent contamination from groundwater aquifers became major societal goals (Soesilo and Wilson, 1997).

In the past pump-and-treat was the technology of choice to remediate most contaminated groundwater. However, it has been recognized that this method are not any more very effective to remediate sites contaminated with chlorinated solvents (Brusseau et al., 2001). Many factors contributed to its ineffectiveness, one of the most common being zones of high concentration or large mass of contaminants in the source zone. While this method removes diluted portion of the plume, contaminant removal from the source zone is usually limited by equilibrium (e.g., solubility) and kinetics (e.g., dissolution, desorption). During the last decades natural attenuation also known as intrinsic bioremediation has been recognized as a promising alternative to pump-and-treat technologies. Natural attenuation is a strategy of allowing natural process to reduce contaminant concentrations over time (National Research Council, 2000). This process involves physical, chemical and biological process, which act to reduce the mass, toxicity, and mobility of subsurface contamination. When the natural attenuation as a remedy is not sufficient, enhanced in-situ biodegradation is commonly employed (Norris and Matthews, 1994). The use of natural attenuation as a remedy for petroleum hydrocarbons is well established (National Research Council, 2000). More recently, natural attenuation has been proposed for chlorinated solvents, and few studies have successfully tested its feasibility as remedy (Clement et al., 2000; Ling and Rifai, 2007). While dispersion, dilution, sorption , and chemical transformation, contribute for natural attenuation process microbially mediated aerobic and anaerobic degradation comprises the major processes for reduction of contaminant mass in the subsurface (Azadpour Keeley et al., 2001). Extensive research on biotic reactions showed that most chlorinated ethenes are subject to a variety of microbial degradation processes that include reductive dechloriantion, aerobic oxidation, anaerobic oxidation and aerobic cometabolism (Chapelle, 2001). The challenge is then to determine that the transformation processes are taking place at a rate that is protective of human health and the environment. An overview of the different microbial degradation pathways are discussed in the sub-sequent sections.

Table 1.1: Summary of physical and chemical properties of selected chlorinated solvents at 25 $^\circ$C (Source: Morrison et al. (2010)).

Compound	Molecular Weight	Vapor Pressure (p°, torr)	Solubility (mg/L)	Henry's constant (H,atm-m^3/mol)	Liquid Density	K_{ow}	Boiling Point ($^\circ$C)
PCE	165.8	18.9	200	0.0174	1.12	400	121.4
TCE	131.5	75	1100	0.00937	1.35	200	86.7
t-1,2-DCE	97	315	6300	0.00916	1.97		48
cis-1,2-DCE	97	205	3500	0.00374	1.63		60

Kow : Octanol-water partitioning coefficient

Degradation Mechanisms

CAHs can be transformed in the subsurface by range of chemically (abiotic) and Biologically (biotic) reactions (Vogel et al., 1987). The mostly occurring abiotic processes under either aerobic or anaerobic conditions are hydrolysis and dehydrohalogenation. The abiotic transformations mostly result in only a partial transformation of compounds and may lead to the formation of a new compound that are readily or less readily biodegraded by microorganisms (Norris and Matthews, 1994). The scope of the degradation mechanisms discussed here is limited to biotic reactions. According to Wiedemeier et al. (1999) biodegradation of organic contaminants in groundwater occurs via three mechanisms: (i) by using the organic compound as the primary substrate for growth (electron donor); (ii) by using the organic compound as an electron acceptor; and (iii) by co-metabolism. When the contaminant is utilized as a primary substrate, it is a source of energy and carbon for the microorganisms. The first two biodegradation mechanisms involve the microbial transfer of electrons from electron donors to electron acceptors. These processes can occur both in the aerobic and anaerobic conditions. Natural organic matter, fuel hydrocarbons and less chlorinated compounds such as VC serve as electron donor. Under aerobic conditions dissolved oxygen, is used as terminal electron acceptor during aerobic respiration. When oxygen is depleted in advancing contaminant plume, anaerobic conditions can develop and lead to the formation of as many as five different down-gradient redox zones, each with a different terminal electron acceptor (Semprini et al., 1995a). Common anaerobic electron acceptors and the associated microbial process, in the order of their redox potential are nitrate (denitrification); Mn (IV) (manganese reduction); Fe (III) (iron reduction); sulphate (sulphate reduction); and carbon dioxide (methanogenesis). Figure 1.1 shows the redox zones of a typical petroleum plume in an aerobic aquifer, showing the progression from the source area to the edge of the plume. The dominant redox reaction will determine the type of bacteria that typically will exist in particular zone and determine the CAH biodegradation mechanisms that may occur. The third biodegradation mechanisms are brought by a fortuitous reaction where an enzyme produced during unrelated reaction degrades the organic compound. Some pollutants, especially the highly oxidized chlorinated solvents, are not amenable to use as primary growth substrate (U.S.Epa, 1989). The oxidation state of carbon in PCE, TCE, DCE, VC and Ethene is respectively; +2, +1, 0, -1 and -2. An overview of the different microbial and abiotic pathways involved in CAH degradation is provided the sub-sequent sections.

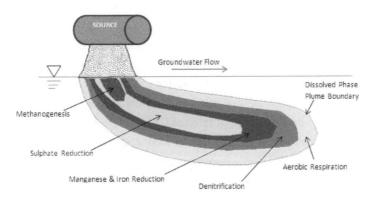

Figure 1.2: Redox zones of a typical Petroleum in an Aerobic Aquifer (modified from U.S.EPA (2000)).

Microbial degradation of CAHs under aerobic conditions

Among the chlorinated ethenes, VC is the most susceptible to aerobic biodegradation, and PCE the least (Norris and Matthews, 1994). In general, the highly chlorinated ethenes (e.g., PCE and TCE) are not likely to serve as electron donors or substrates for microbial degradation reactions (Chapelle, 2001). The tendency of chlorinated ethenes to undergo oxidation increases as the number of chlorine atom present in the compound decreases (Vogel et al., 1987). Norris and Matthews (1994) suggested that both DCE and VC degrade via direct or co metabolic aerobic pathways. TCE can be aerobically transformed through cometabolism or fortuitous oxidation, primarily by organisms that contain an enzymes that are designed for other purpose such as those initiating oxidation of hydrocarbons or ammonia (McCarty and Criddle, 2012; Reineke et al., 2002). Aerobic biodegradation is rapid compared to the other mechanisms of VC degradation, especially reductive dehalogenation (Wiedemeier et al., 1999). According to Chapelle (2001) since production of DCE and VC occurs by reductive dechlorination under anaerobic conditions, the aerobic oxidation of these compounds is often limited in groundwater systems. However, in areas where anaerobic condition that produce DCE and VC change to more oxic conditions, which often happens on the fringes of contaminant plumes, aerobic oxidation of these compounds can be significant. Moreover, aerobic degradation of DCE and VC can be stimulated in groundwater systems by providing a source of oxygen.

Microbial degradation of CAHs under anaerobic conditions

In anaerobic condition, when neither oxygen nor nitrate is present, PCE and TCE can be used by certain microorganisms as electron acceptors in energy metabolism (McCarty and Criddle, 2012). The process is a step wise reaction in which one chlorine is removed at time and replaced by hydrogen. This process is known as reductive dehalogenation or reductive dechlorination. As an example, PCE can be reductively dechlorianted to TCE, which in turn can be reduced anaerobically to cis-dichloroethene (cis-DCE), which can be converted to vinyl chloride (VC) and ethene. Biodegradable organic materials must be present as electron donors for reductive dechlorination of chlorinated solvent to occur. Table 1.2 presents the literature values of first order degradation rate for reductive dechlorination of selected CAHs. More detailed report of aerobic and anaerobic degradation rate constants under laboratory and field conditions is available in Schaerlaekens et al. (1999). Anaerobic oxidation DCE and VC is also reported in literature. It was shown that VC, and to a lesser extent DCE, could be oxidized to carbon dioxide under Fe (III)-reducing conditions (Bradley and Chapelle, 1996). Biodegradable organic materials must be present as electron donors for reductive dechlorination to occur. The electron donor that appears to be most preferred by dehalogenating organism is H2 (McCarty and Criddle, 2012).

Table 1.2: First-order biodegradation rate coefficients for reductive dechlorination of CAHs.

Compound	λ (day^{-1})	λ (day^{-1})
PCE	0-0.080	0-0.40
TCE	0-0.023	0-3.13
Cis-1,2-DCE	0-0.130	0.001-0.200
DCE (other isomers)	0.001-0.006	0.010-0.270
Vinyl chloride	0-0.007	0-5.20

Source Surampalli (2004)

1.1.2   Nitrate, groundwater, and receiving water quality

Groundwater contamination by nitrate is a globally growing problem. There are many local sources of nitrate that contribute to groundwater quality problems. However, many researchers have shown that the main source of nitrate delivery to groundwater and surface water is from agriculture. The greatest problems with nitrate arise from heavy fertilization in agricultural fields. Numerous studies have shown that nitrate concentrations in shallow groundwater can be directly linked to agricultural land use. Increased nutrient levels in surface streams and eutrophication of some coastal plain waters has led to control of nitrate losses from agricultural fields (Jacobs and Gilliam, 1985).

Currently there is a growing interest on riparian zones, regarding the protection of the streams against agricultural pollution (Bren, 1993). Riparian zones are the most important elements of the hydrological landscape. These near stream zones have become a focus of attention because they are hot spots for biodiversity, and their role in nutrient removal. In riparian wetlands and floodplains complex surface water and shallow groundwater interaction exists. These areas are typical landscape features in many lowland areas. The positions of the riparian wetlands in the landscape make them important zones for nutrient retention and transformation before nutrients are transported directly to streams or rivers. The high water table and accumulation of organic matter provided by vegetation enable these zones to remove nitrate by denitrifiction before it could enter streams (Jacobs and Gilliam, 1985). Important controlling factors for riparian zones for removing nitrate include: the rate of inflow of

N-rich water into the riparian zone, the residence time of the water, how long it remains in the riparian zone, the concentration of organic matter, and the available surface area of plants and other substrates for growth of microbes (Woltemade, 2000). Accurate characterization of the nutrient balance in the riparian wetlands is only possible with understanding of the GW-SW dynamics, the residence time of groundwater flow in the riparian wetland, the rate of mineralization and denitrification. When the water table in the riparian wetland is artificially lowered for any reasons (e.g., channelization, land drainage, abstraction) this will have significant effect on the function of the riparian wetland.

Drainage measures in riparian zones result in loss of ecosystem service that riparian zones can provide (Zedler, 2003). One of these services is nutrient removal capacity. Drainage systems allows a faster removal of groundwater from the riparian zone, hence, the high potential for nutrient retention and modification of riparian wetlands and particularly the retention of nitrate is reduced (Schmalz et al., 2009). In addition, these drainage systems significantly affect the discharge pathway in the riparian wetland and their hydrologic integration in the landscape. Drainage results in faster groundwater flow and impairs denitrification and plant uptake (Burt, 1995). When large amounts of water bypass the saturated riparian soils, or flows deep below the flood plain alluvium, the buffer zone will be ineffective (Burt and Pinay, 2005). However, the services of riparian wetlands can be restored through careful planning and designing of drainage systems.

1.1.3   Excessive pumping in coastal aquifers

Over-exploitation of aquifers that are in contact with the sea or other water bodies of inferior quality may result in the deterioration of the aquifer's water quality. Contaminants from surface water may reach pumping wells in the aquifer and hinder further supply from the aquifer. These could significantly reduce the aquifer potential to provide freshwater. In coastal areas, over pumping may induce landward movement of the freshwater-saltwater interface, causing seawater intrusion. Seawater intrusion is an important problem in coastal areas where large quantities of freshwater are needed for domestic use and agriculture. The quantity of water available from an aquifer depends basically upon the amount of natural recharge, the rate at which groundwater is abstracted and the available underground storage. With the increasing development of groundwater resources and the growing impacts of human activities on the aquifer, problems such as groundwater level decline and groundwater quality deterioration would continue to be a great concern.

In many arid and semi arid countries, potable water supplies are pumped out of the groundwater. As the groundwater wells pump more water from the aquifer to cater the domestic need of an increasing population, seawater from the sea is moving towards the aquifer to replace the lost amount. A progressive decline in groundwater levels and seawater intrusions are an indication of future management problems due to the fact that they are consequences of exceeding the safe yield of the aquifer. Various methods available for preventing seawater intrusion into coastal aquifers include: reduction of abstraction rates, relocation of abstraction wells, subsurface barriers, natural recharge, artificial recharge and abstraction of saline water. Since fresh water resources in many semi-arid and arid countries are scarce, artificial recharge using treated municipal wastewater is a preferred groundwater management intervention that could help to improve groundwater level decline and control seawater intrusion.

In summary, the stresses on groundwater, both in terms of quantity and quality are growing rapidly. Groundwater contamination is an issue which has received special attention, because it results in poor drinking water quality, loss of water supply, degraded surface water systems, high cleanup cost and potential health problems. To address these issues we need to understand the response of aquifer systems to contamination and the interactions between groundwater and surface water. Modelling provides powerful tools for such understanding and evaluations.

1.2   Using Models to solve groundwater problems

1.2.1   Groundwater Flow Models

Groundwater flow models are physically based mathematical models derived from Darcy's law and law of conservations of mass (Konikow and Mercer, 1988). Numerical methods, Finite-Difference, Finite-Elements, and Finite-Volume methods are widely used for solving the groundwater flow equation provided that model parameters, initial and boundary conditions are properly specified. Groundwater flow models are routinely employed in making environmental resources management decisions (ASTM, 2010). They are tools that can aid in studying groundwater problems and can help to increase our understanding of groundwater systems (Mercer and Faust, 1980). According to Zhou and Li (2011) groundwater flow models can be used: (1) for investigating groundwater dynamics and understanding the flow patterns; (2) for analyzing responses of the groundwater systems to stresses; (3) for evaluating recharge, discharge and aquifer storage processes, and for quantifying sustainable yield; (4) for predicting future conditions or impacts of human activities; (5) for planning field data collection and designing practical solutions; (6) for assessing alternative policies; and (8) for communicating key messages to public and decision makers. Groundwater flow models are also the main component of contaminant transport models. They are used to compute the velocity fields, which are required by the transport models.

1.2.2 Contaminant Transport Models

Numerous mechanisms control the movement of contaminants through porous media. The transport processes associated with even the simplest forms of contamination, conservative, non-reactive dissolved constituents involve: (1) advection, which describes the transport of solutes by the bulk motion of flowing groundwater and (2) hydrodynamic dispersion, which describes the spread of solutes along and transverse to the direction of flow resulting from both mechanical mixing and molecular diffusion. These processes have been described quantitatively by a partial differential equation referred to as the advection-dispersion solute transport equation. Contaminant transport models are used to solve the advection-dispersion equation and to yield the contaminant concentration as a function of time and distance from the contaminant source. If the contaminant is reactive, terms describing chemical reactions, reactions mediated by microorganisms or caused by interaction with the aquifer material are added to the advection-dispersion transport equation.

According to Aral and Taylor (2011), contaminant transport models are used for: (1) predicting the fate and transport of contaminants in the subsurface, (2) assessing the potential for natural attenuation at contamination sites, (3) designing long-term monitoring systems in order to ensure that adverse contaminant migration is detected, (4) designing engineering remediation strategies , (5) determining the capture zones for a specific well and pumping rate arrangement in the pump-and-treat remediation strategies, (6) estimating the time required for cleanup to be accomplished, (7) analyzing the fate and transport of contaminants microorganisms, and associated constitutes for the engineered bioremediation systems, (8) evaluating other remediation approaches such as natural attenuation, reactive barrier etc. The aim of contaminant transport analysis in a planning context is to determine the concentration and travel times of potential pollutants at critical discharge points, such as springs, wells, or surface water bodies (Page, 1987) .

1.2.3 Simulation-Optimisation models

Models which solve the governing groundwater flow and solute transport equations (simulation models) in conjunction with an optimisation technique (optimisation model) have been increasingly used as aquifer management tools. These models are called simulation-optimisation models or groundwater management models. When simulation models are used alone, developing a good aquifer management option to address a particular problem often requires time consuming and iterative trial-and-error approach. In using trial and error, the obtained results are also to certain extent subjective. Use of such approach often fails to consider important physical and operational restrictions (Gorelick, 1983). Optimisation approaches entail mathematical definition of the problem to be solved, using decision variables, state variables, objective functions, constraints, and bounds (Pereira et al., 2002). Decision variables are the unknown quantities that need to be managed. State variables are quantities that define the state of the system (e.g. head or concentration). Objective function is a goal that is minimized or maximized during optimisation to achieve the most important modelling goal (e.g., cost, total pumping rate). Constraints are restrictions that must be obeyed in the final design (e.g., drawdowns, head, gradients, or concentrations).

Depending on the choices of decision variables, constraints, and objective functions, simulation-optimisation models have been used for a range of groundwater management problems. For example, aquifer remediation problems (Ahlfeld, 1990; Kuo et al., 1992; Maskey et al., 2002) conjunctive use of surface and groundwater (Rao et al., 2004; Reichard, 1995; Safavi et al., 2010) controlling seawater intrusion (Abarca et al., 2006; Reichard and Johnson, 2005) artificial groundwater recharge studies (Jonoski et al., 1997). Many other applications of groundwater management problems are also possible. Reviews of simulation-optimisation approaches for groundwater management problems (although not very recent) are available in Gorelick (1983), Wagner (1995).

1.3   Problem Statement

Industrial activities are often located nearby rivers to facilitate transport operations of industrial manufacturing. Because of this, numerous contaminated sites exist close to rivers. Contaminants emerging from old industrial activities are posing major risks to the environment, particularly due to discharge of contaminated groundwater to the surface water bodies. In order to redevelop these areas for new economic activities or other uses, and to protect the receiving water body from such contamination, remediating these sites is necessary. For remediation, many site-specific factors have to be considered, among which the potential for natural attenuation, dynamics of groundwater-surface water interactions and biogeochemical processes.

GW-SW interactions provide one of the major pathways through which chlorinated solvents or other groundwater contaminants interact with humans and the wider terrestrial environment. Water fluxes between groundwater and surface water mediate the transport of contaminants between these two hydrologic compartments. Without these fluxes, the hydrologic cycle is not complete and the surface processes are not linked to the subsurface processes (Gunduz, 2006). Understanding the interaction between groundwater and surface water is relevant for many purposes including aquifer remediation, wetland management, controlling seawater intrusion and conjunctive use of groundwater and surface water. For example, for areas where shallow groundwater is polluted by nitrate, quantifying GW-SW interaction between the shallow aquifer and surface water bodies will help to address the effect of polluted aquifers in surface water bodies and consequently on the environment (Essaid et al., 2008).

Physical and chemical properties of surface water and groundwater are different. When they mix physicochemical gradients are usually established (Brunke and Gonser, 1997). These gradients are highly dependent on the flow patterns as they determine the extent of mixing. Past studies also showed that the attenuation capacities of streambed sediments are strongly dependent on these fluxes exchanges. For example, Chapman et al. (2007) reported that TCE plume attenuation is strongly depend on GW-SW interaction. In similar study Puckett and Hughes (2005) reported the same for agricultural chemicals. On the contrary, the GW-SW fluxes exchange in the hyporheic zones are complex, spatially, and temporally varying (Gunduz and Aral, 2005). One of the biggest challenges for modelling is the significant time scale difference between the surface and groundwater processes (Rassam et al., 2011). Modelling GW-SW interaction in the hyporheic zone with regional groundwater and transport model is very challenging because: (1) the spatial scale to which the model is applied is too big to account such small scale processes, (2) calibrating transient regional groundwater and contaminant transport in high temporal resolution is not possible due to limitation of data. Therefore, modelling approaches that can resolve the spatial patterns of GW-SW interaction on a meter or sub-meter scale on a given stream reach and temporal scale that enable to capture important details of hydrologic events are essential.

Riparian zones also known as riparian wetlands exist along a river or stream, and act as a transition zone between the terrestrial and the aquatic ecosystems (Burt et al., 2010). They have a high potential to regulate nutrient fluxes between the upland areas and the stream (Cey et al., 1999; Warwick and Hill, 1988). The groundwater in these areas mostly exists at shallow depth and vegetation and soil processes may therefore modify the chemistry of water before entering the stream (Lowrance et al., 1985; Warwick and Hill, 1988). Construction of artificial drainage systems such as open drainage ditches or tile drains are common in these areas because riparian zone conditions results in rich organic soils that are much needed for agriculture (Winter et al., 1998). These drainage systems considerably affect the proper function of riparian zones and increase nutrient levels entering to streams and rivers. For instance, Winter et al. (1998) have shown that nitrate removals in riparian zones are significant where a large fraction of the local groundwater flow is passing through the reducing sediments, which are present at shallow depth. This is because the microbial denitrification processes. The main process by which riparian zones remove nitrate is reported to be significant at location where oxygen is absent or present at very low concentration. This means that the groundwater level should be high enough to reach to sediments containing high organic carbon to create this anaerobic condition. If the groundwater is flowing deep the denitrification processes will be slow. On the other hand, if the groundwater flows through the riparian zones at high velocity, it may not have a long enough residence time for plant uptake, through which denitrification occurs. Therefore understanding riparian zone hydrology such as flow paths and residence time, groundwater-surface water interactions, and shallow groundwater level management is very important because these conditions determine the rate of nutrients removal in drained riparian zones. The term drained riparian wetland is used throughout this thesis to represent riparian zones that are drained with artificial open drainage ditches.

Many arid and semi-arid regions rely on groundwater for their domestic supply. Increasing demands and climate change created new challenge for water resource managers in these regions. Furthermore, depletion of costal aquifers due to increased demand further increased risk of seawater intrusion. Understanding and responding to these challenges requires taking into account the hydrologic conditions and myriad of groundwater management alternatives. The developments of any successful groundwater management alternatives often require consideration of different management objectives and set of constraints. For instance, efficient management strategies are required for optimal use of groundwater from costal aquifer without further intrusion of seawater and simultaneously meeting the required demand, or efficient management are needed to artificially recharge the aquifer for aquifer storage and recovery, and simultaneously preventing seawater intrusion. Optimal management strategies cannot be identified using simulation models alone. Simulation-optimisation approaches that explicitly account the management objectives and set of constraints in the modelling process are required.

1.4 Research Objective

As groundwater resources come under increasing pressure due to contamination and increasing demand, it is important to adopt modelling approaches and develop tools that can solve groundwater problems and lead to better management. The overall objectives of this thesis are: (a) to investigate and highlight the value of integrated modelling approaches; (b) to asses and evaluate different aspects of complex groundwater resource problems; (c) to explore the feasibility of a combined simulation-optimisation approach. More specifically, these objectives can be formulated as: (i) how can groundwater flow and contaminant transport models improve our understanding of contaminant transport? (ii) how do the spatial and temporal scales influence groundwater-surface water interactions? (iii) how can we improve groundwater management using an integrated simulation-optimisation approach?

The approach followed in this study to explore the concept of integrated modelling led to the following component studies: (i) developing a groundwater flow and contaminant transport model; (ii) integrating different modelling techniques to investigate groundwater-surface water interaction at different spatial and temporal scales; and (iii) integrating groundwater models with different optimisation techniques. These approaches demonstrate the potential combination of tools that could be used to investigate the different groundwater resource problems described in the previous sections. Furthermore, the integrated modelling approach developed in this study can extend the scope and use of the standard groundwater flow model MODFLOW. From specific case studies, the capabilities of integrated modelling approaches for solving groundwater resource problems are investigated. The specific research questions can be formulated as:

1. Assess the feasibility of natural attenuation as remedial option for chlorinated solvent contaminated groundwater.
2. Investigate the effect of temporal resolution on local scale groundwater surface water interactions using combined groundwater flow and heat transport models.
3. Quantify groundwater-surface water interaction in drained riparian wetlands to obtain insight into the dynamics of exchange patterns using numerical and analytical models.
4. Provide shallow groundwater level management strategies in riparian wetlands with artificial drainage by combining groundwater flow models with global optimisation techniques.
5. Assess the feasibility of managed aquifer recharge in arid coastal zones by combining groundwater flow models with multi-objective optimisation techniques.

1.5 Dissertation Structure

The thesis is organized in eight chapters. A brief overview of the structure is as follows:

Chapter 2 presents literature review. It describes the governing groundwater flow equation and numerical solutions, provides an overview of the main contaminant transport processes, and describes the governing transport equation. It also provides an overview of groundwater-surface water interaction, scaling issues in groundwater-surface water interactions, methods of quantifying groundwater-surface water interactions. It also describes the models used in groundwater-surface water interaction and mathematical optimisation techniques. The objective of this chapter is to provide theoretical background required for field application case studies in the consecutive chapters.

Chapter 3 presents a modelling case study to remediate groundwater systems contaminated with chlorinated solvents. After a short description of the study area, it presents the modelling framework; groundwater model calibration, contaminant transport model description and calibration. Finally, natural attenuation scenario, source removal scenario and sensitivity analysis results are presented.

Chapter 4 presents local scale GW-SW interaction modelling case study using temperature as natural tracer. It begins with the description of the case study area, followed by the description of the temperature and water level data used in the modelling, the modelling approach, model calibration and validation. Finally it presents the simulation results.

Chapter 5 presents GW-SW interaction in drained riparian wetland using numerical and analytical modelling approaches. Firstly, it begins with the description of the study area and collected data. Secondly, it describes the numerical model design. Thirdly, it presents the analytical modelling approach. Finally, results obtained using both numerical and analytical models are presented.

Chapter 6 uses the numerical simulation model built in chapter 5 and couples it with optimisation algorithms for optimal management of shallow groundwater level in the drained riparian wetland. Firstly, it presents the optimisation problem formulation. Secondly, it describes the software used. Thirdly, it describes the optimisation algorithms. Finally, it presents the simulation-optimisation results.

Chapter 7 presents simulation-optimisation model application case study for assessing the feasibility of managed aquifer recharge. Firstly, a review of simulation-optimisation approaches used in groundwater management problems described.   Secondly, description of the study area is provided. Thirdly, the numerical model design and calibration process are described. Fourthly, the optimisation problem formulation is described. Finally, results are presented and discussed.

Chapter 8 summarizes the conclusions of the research based on the various case studies presented in this thesis and suggests recommendations for further research.

2   Literature Review

2.1   Introduction

This chapter describes some of the important concepts in groundwater and contaminant transport modelling. Topics covered in this chapter include solute transport processes, groundwater and contaminant transport modelling, governing equations in groundwater and contaminant transport modelling and solution methods, NonAqueous Phase Liquids, NAPL dissolution, groundwater-surface water interactions, and mathematical optimisation

2.2   Solute Transport Processes

Numerous mechanisms control the movement of contaminants through porous media. The transport processes associated with even the simplest forms of contamination, conservative, non-reactive dissolved constitutes, involve advection, dispersion, and molecular diffusion. A brief review of processes and how those processes can be represented in the advection-dispersion equation is described.

2.2.1   Advection

Advection also referred as convection is a primary process by which solute is transported by the average linear groundwater velocity. In a saturated medium, this velocity can be calculated using Darcy's Law. As the groundwater moves it picks up the contaminant and moves it in the direction of groundwater velocity. The amount of contaminant that the water can carry is controlled by the solubility of that particular compound.

2.2.2   Dispersion

Dispersion is a process in which dissolved contaminants are spread as they move with the groundwater as results of two basic process, molecular diffusion and mechanical mixing; Mechanical mixing results from variations in groundwater velocity within the porous aquifer caused by aquifer heterogeneity that cause fluctuations in the local flow directions relative to the mean flow direction. Molecular diffusion results from the random collision of solute molecules and produces a flux of solute particles from areas of higher to lower solute concentration (Bear 1979). In other words, the contaminants move from the high concentration to the low concentration areas to form a uniform concentration distribution.

Dispersion and spreading during transport result in the dilution of contaminant and the attenuation of concentration peaks. Dispersive spreading may also results in the arrival of detectable contaminant concentration at a given location significantly before the arrival time that is expected solely on the basis of the average groundwater flow rate (Mackay et al., 1985). Mechanical dispersion plays a major role in the total dispersion. However, when the flow velocity is extremely low, molecular diffusion may become prominent. Obviously, both mechanical dispersion and molecular diffusion will force the contaminant to move longitudinally and transversally with respect to the mean flow direction. In a typical laboratory test, one-dimensional flow is created in a homogeneous porous column and solute at constant concentration is introduced at the column entrance. By measuring the concentration as a function of time in the effluent and matching it with the standard solution of the diffusion equation, one can determine the longitudinal dispersion coefficient and subsequently the dispersivity (Freeze and Cherry, 1977).

2.2.3   Sorption and retardation

Sorption is a physical process in which solutes are absorbed into the solid matrix. As the result of sorption the transport of solute relative to the bulk groundwater is slowed down. This phenomenon is usually referred as retardation. Sorption and desorption are complex processes and are a function of the type of contaminant, the composition of the solid, the chemistry of the aqueous phases and the distribution of contaminant between the solid and fluid phases (Hall and Johnson, 1992). The higher the fraction of the contaminant sorbed, the more retarded is its transport. The degree of sorption depends primarily upon the organic carbon content of the soil and the type of compound. Sorption causes contaminants to move slower than the flowing groundwater. Sorption increase the time required to remediate contaminated aquifer. An almost linear relation exists between retardation and time of remediation for a specific level of contaminant (Konikow and Mercer, 1988).

A number of conceptual and empirical models have been developed to describe various adsorption patterns. The most simple is the linear model, which describe the accumulation of solute by the sorbent as directly proportional to the solution phase concentration ($Cs=Kd*Caq$). Linear approximations to sorption equilibrium data are particularly useful in modelling contaminant fate and transport because they substantially reduce the mathematical complexity of the modelling effort (Weber et al., 1991). The Langmuir model and Freundlich isotherm model are the most widely used nonlinear sorption equilibrium models. A comprehensive review of sorption phenomena in subsurface system is presented by Weber et al (1991). Because of its simplicity and ease of implementation in analytical or numerical models, the local equilibrium assumption is very attractive for application to two and three dimensional flow model. It is believed that because of the slow movement of groundwater in natural system, an assumption of local equilibrium between the soil grains and the pore fluid was a reasonable one (Abriola, 1987).

2.2.4   Reactions

Under reactions we commonly understand chemical, biological, and physical reactions between groundwater and solid aquifer material. Chemical reactions may include redox reactions, acid-base reactions, precipitation-dissolution, or other complex reactions depending on the compounds. Naturally occurring microbes in the soil and aquifer may be able to biodegrade many organic compounds if the level of contamination is low and does not produce toxicity for the active bacteria. However, in many cases it is possible to enhance the ability of naturally occurring bacteria to degrade organic contaminants by the addition of nutrients, oxygen and water.

2.3   Contaminant Properties Affecting Transport

The type, extent and duration of organic chemical contaminations of groundwater are controlled by the properties of contaminants, geochemical, physical and biological processes in the ground and hydrogeological conditions. This section presents the properties of contaminant, their interactions within groundwater, difficulties and concerns associated with their presence in the subsurface.

Contaminant behaviour in the subsurface is affected by its own physical and chemical characteristics as well as the natural characteristics of the subsurface (Soesilo and Wilson, 1997). Some of these properties are solubility, volatility, tendency to adsorb to solids, chemical reactivity, biodegradability, and density. From the perspective of groundwater pollution, the most significant contaminant characteristic is solubility (Gorelick et al., 1993). Solubility defines the maximum possible concentration that commonly occurs in groundwater for a given contaminant. It depends on chemical structures (molecular size, functional groups, freedom of rotation, physical state), solvent (PH, ionic strength, polarity) and environmental conditions (temperature, pressure). This chemical property, perhaps more than any other property, affects the fate and transport behaviour of organic chemicals in groundwater. Liquids that are immiscible with water are of special interest in groundwater flow problems. The failure of such liquids to completely dissolve in groundwater produces stratified flow problem. Low-density immiscible liquids also referred to as Light NonAqueous Phase Liquid

19

(LNAPLs) float on the surface of groundwater. On the other hand, high density liquids, referred to as Dense NonAqueous Phase Liquid (DNAPLs) sink through the groundwater until they reach the aquifer bottom. BTEX (benzene, toluene, ethylene, and xylene) is an example of LNAPL, and PCE and TCE, known to be major causes of soil and groundwater contamination, are representative substances of DNAPLs (Kamon et al., 2004).

One of the biggest challenges to overcome in the field of soil and ground water remediation is the problem of cleaning up sites contaminated with DNAPLs. It can be quite difficult to locate and recover DNAPLs once they are released into the subsurface environment (Interstate Technology et al., 2000). When spilled, the DNAPL will travel vertically through the subsurface due to gravity. In general, the behaviour of DNAPL in the subsurface is very complex, depending on the spill history, DNAPL properties and geologic heterogeneity. This renders difficult the efforts to locate the DNAPL in the subsurface and the subsequent remediation of these sites (Dou et al., 2008).

2.4   Nonaqueous Phase Liquids

Many contaminant including chlorinated solvents enter the subsurface in the form of Nonaqueous Phase Liquids (NAPLs). NAPLs are immiscible fluids that do not readily mix with water and therefore flow as a separate phase from the groundwater. Migration of NAPLs is a complex process. NAPLs generally migrate downward through the vadose zone due to gravitational force. If the spill (or leak) is sufficiently large and the NAPL is less dense than water (known as LNAPL, e.g. petroleum oil), it will eventually reach the water table, where it laterally spreads in the capillary fringe zone. On the other hand, if the NAPL is heavier than water (known as DNAPL, e.g. PCE, TCE), it continues to migrate by displacing aquifer pore water (Mackey et al 1985, Powers 1991). During the migration process, a portion of NAPL may also be trapped as discontinuous blobs in pore space within the saturated zone. They can also pool over low permeable zones. Figure 2.1 shows the infiltration behaviour of NAPLs. One of the biggest challenges to overcome in the field of soil and ground water remediation is the problem of cleaning up contaminated sites with NAPLs. Remediation problem is becoming even more complex when the spill is related to DNAPLs. It can be quite difficult to locate and recover DNAPLs once they are released into the subsurface environment (Suthersan, 2010). DNAPLs pools can serve as a long-term source of dissolved contaminant plumes.

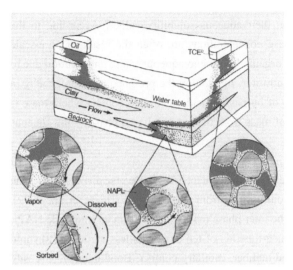

Figure 2.1 : Comparison of the infiltration and spreading behaviour of two hydrophobic substances in a porous medium (Source: Mackay and Cherry (1989)).

2.5 NAPL Dissolution: Equilibrium and Rate-Limited Behaviour

The evolution of dissolved plumes from the DNAPL contaminated zone (with either pools or blobs) depends on the DNAPL solubility and mass transfer characteristics of DNAPL-water interface (Clement, 1998). In the aquifers, dissolution occurs when groundwater dissolves residual components of DNAPL blobs or a large continuous DNAPL pool. If all the steps of the mass transfer process are very fast compared with the advective and dispersive transport in the aqueous phase, the water in the vicinity of the organic phase will always be in equilibrium with the organic phase. This is known as the Local Equilibrium assumption or Raoult's Law Approach (Heyse et al., 2002). The Raoult's Law states that the equilibrium aqueous concentration of a compound in a NAPL mixture is the product of the mole fraction of the species in the mixture and the single component aqueous solubility (Brusseau et al., 2001). This law is based on the assumption that the organic phase activity coefficients are equal to unity; that is, the mixture is assumed to be ideal. The water originating from the region of entrapped NAPL is assumed to be at maximum aqueous concentration (Okeson et al., 1995) Several investigators have studied the characteristics of DNAPL dissolution process in saturated porous media (Clement et al., 2004a; Clement et al., 2004b; Geller and Hunt, 1993; Hunt et al., 1988; Imhoff et al., 1994; Miller et al., 1990; Powers et al., 1994; Powers et al., 1991). They have shown that the limitations on mass transfer between the phases may prevent the system from establishing equilibrium aqueous concentrations. Powers et al. (1991) reported that there is clear evidence from filed data that concentration of organic contaminants in groundwater at the contamination site containing NAPLs are lower than their aqueous solubility values. According to the authors these lower dissolved phase concentrations are often the result of pore-scale mass transfer limitations from the organic liquid into aqueous phase and dilution due to mixing with uncontaminated groundwater. In addition the presence of heterogeneity in the source zone may create zones of by passing and /or limited contact where there is little opportunity for the transfer of mass from the organic liquid phase to the aqueous phase.

The rate of NAPL dissolution depends on the contact area, transverse dispersion and the rate at which water moves the solute away from the interface (Heyse et al., 2002). According to Miller et al. (1990) the inter phase mass transfer from an immobile NAPL phase to a mobile aqueous phase is a function of ten dimensionless variables (Reynold number, Schmidt number, Serwood number, capillary number, Bond number, viscosity mobility ratio, Goucher number, Peclet number, Stanton number, and Weber number), which are in turn a function of several other parameters (dynamic viscosity and density of the aqueous and NAPL phase, diameter of the porous media particle, mean pore velocity of the aqueous phase, molecular diffusion of the NAPL source solute in the aqueous phase, the interfacial tension between the NAPL and aqueous phases, and the acceleration due to gravity). Powers et al. (1994) proposed a model for transient dissolution rates in terms of porous medium properties, Reynolds number, and volumetric fraction of NAPL. This model was developed based on one dimensional experiment conducted using homogenous sandy soils, hence may not be applicable for heterogeneous multi-dimensional flow domains. Some studies are also studied the dissolution kinetic of individual NAPL blobs using a glass bed micro-model (Kennedy and Lennox, 1997). They determined the micro-scale mass transfer coefficient of individual blobs by measuring the temporal change of the volume and surface area of spherical and ellipsoidal blobs using image analyzer. Other studies e.g., Clement et al. (2004b) and Haest et al. (2012) have used sand box laboratory experiment to determine the DNAPL dissolution rate.

In a practical filed problems estimating inter phase mass transfer rate is difficult, if not impossible; hence if NAPL source is known or suspected to be in the system, a partitioning relationship is often use to model the source region having concentrations close to the saturation (Schäfer and Therrien, 1995; Touchant et al., 2006). Clement et al. (2000) used inverse modelling to estimate the source loading rate of PCE and TCE at Dover site. The authors used the field estimated mass of each contaminant species to determine the mass release rate of the parent compounds using model calibration.

2.6   Modelling Groundwater Flow and Contaminant Transport

Groundwater flow and contaminant transport modelling is an important component of most aquifer remediation studies. Designing an effective and efficient system for the control of contaminated groundwater requires an understanding of the contaminant distribution and transport mechanism with an aquifer. It is also necessary to carry out a quantitative evaluation of the impact that alternative capture system design will have on contaminant transport. In practice, both tasks are almost always attempted through the use of models of the groundwater systems (Gorelick et al., 1993). The use of groundwater modelling for contaminant transport predictions are also common in the risk-based decision making process. Groundwater flow models are used to calculate the rate and direction of movement of groundwater through aquifer whereas contaminant transport models simulate the movement and chemical alteration of contaminant as they move with groundwater through the subsurface. Models are generally used to support remedial decisions where groundwater contamination exists above a prescribed action level.

2.6.1   Groundwater Flow Modelling

Groundwater flow system is dynamically linked to the hydrological cycle through various natural or artificial recharge processes. As part of the hydrological cycle, groundwater is always in motion from regions of recharge to discharge points such as rivers, lakes, or oceans. The role of a groundwater flow model is to characterize the balance of withdrawal or recharge events so that changes in local groundwater flow rates and changes in water levels can be predicted. Stated in another way, groundwater flow models are used to calculate the rate and direction of groundwater flow through aquifer.

Henry Darcy, French engineer and scientist has performed laboratory experiments 1856 on the flow of water through a sand column and formulated the well known Darcy formula (Darcy, 1983). This formula is the basic law that governs the flow of water through porous media. Darcy's Law states that the groundwater flow rate is proportional to the hydraulic gradient, with a constant of proportionality (hydraulic conductivity) that depends on the character of the porous medium and the fluid. When combined with a mass balance on water, the Darcy formula creates the basis of the groundwater flow model. Darcy's Law has limits on its range of applicability, because it was derived from experiments on laminar flow of water through porous material. Another issue associated with Darcy's Law is that it appears to have a lower limit of applicability and may not be valid in extremely fine-grained material (Konikow and Mercer, 1988). This is because as the velocities are getting very low other driving force other than hydraulic gradient such as temperature gradient, chemical gradient or even electrical gradient, may cause fluid flow that is comparable to that driven by the hydraulic gradient.

2.6.2   Governing Groundwater Flow Equations

Mathematical formulations for groundwater flow are based on the principles of conservation of mass and momentum. Conservation of mass is expressed by the continuity equation. Darcy's law provides an equation of motion which is applied in many saturated flow situations. The governing equations are formulated for each grid location, and the distribution of heads and velocities is computed accordingly. The governing equation applied in most general ground water models (e.g. MODFLOW) has Equation 2.1 forms   (Kresic, 2006; National Research Council, 1990).

$$\frac{\partial}{\partial x}\left(K_{xx}\frac{\partial h}{\partial x}\right) + \frac{\partial}{\partial y}\left(K_{yy}\frac{\partial h}{\partial y}\right) + \frac{\partial}{\partial z}\left(K_{zz}\frac{\partial h}{\partial z}\right) - W = S_s\frac{\partial h}{\partial t} \tag{2.1}$$

Where K_{xx}, K_{yy} and K_{zz} are values of hydraulic conductivity along the x, y, and z coordinate axes, h is the hydraulic head, W is a flux term that accounts for pumping, recharge, or sources and sinks, S_s is the specific storage, x, y, and z are space coordinates, and t is time.

Numerical methods such as Finite-Difference Method (FDM), Finite-Element Method (FEM), and Finite-Volume Method (FVM) are widely used for solving the governing flow equation (Equation 2.1). Numerical methods replace the continuous problem represented by the partial differential equations into a finite set of points or volumes via mesh or grid (Hinkelmann, 2005; Toro, 1997). They transform the partial differential equations to algebraic equations. The governing equations are formulated for each grid, element or volume and the distribution of heads, and velocities are computed.

In the FDM method the computational domain is divided into grids, and the groundwater head are solved at each node or centre of each grid. Taylor series expansions mostly used to obtain approximations to the first and second derivatives of the variables with respect to the coordinates. One of its disadvantage is that the method does not necessary guarantee mass and momentum conservation (Ferziger and Perić, 1996; Hinkelmann, 2005). Also, the restriction to simple geometries is a significant disadvantage in complex flows. The computational domain in the FEM is subdivided into a large number of small elements, in each of which the groundwater head is approximated by some simple function (Bear and Verruijt, 1987). The model domain can be divided into triangular or quadrangular elements of arbitrarily shape. But, triangular elements are preferred options as such elements allow to closely approximate natural (curved) boundaries and allow dense mesh in the region of interest and coarser in areas where the flow is of less interest. FEM is similar to the FVM in many ways. The distinguishing feature of FEM is that the equations are multiplied by weight function before they are integrated over the entire domain (Ferziger and Perić, 1996). Compared to FDM the FEM has two advantages: it guarantees global mass and moment conservation and it has the ability to deal with arbitrarily geometries using unstructured meshes (Hinkelmann, 2005).

FVM are based on the integral form of the conservation law instead of the differential equations (LeVeque, 2002). The solution domain is subdivided into a finite number of contiguous control volumes (CV), and the conservation equations are applied to each CV. The control volume is defined for every node and the differential equations in their conservative form are integrated over all volumes. This methods are reported to be very effective to handle discontinuous solutions (LeVeque, 2002). The FVM can accommodate any type of grid, so it is suitable for complex geometries. FVM is recently introduced into groundwater modelling through MODFLOW-USGS (Unstructured Grids) (Panday et al., 2013). This new version of MODFLOW-USGS allows refined grid on upper layer where rivers and streams exist and coarse refinement in other layers. The disadvantage of FVM compared to FDM is that methods of higher orders than second are more difficult to develop in 3D. This is because the FVM requires three levels of approximation: interpolation, differentiation, and integration (Ferziger and Perić, 1996). It is beyond the scope of this study to give completer description of each method.

In order to obtain a unique solution of the partial differential equation (Equation 2.1) additional information about the physical states of the system such as initial and boundary conditions are needed. The initial conditions are simply the values of the dependent variable specified everywhere inside the boundary. Boundary conditions are mathematical statements specifying the dependent variable (head) or the derivative of the dependent variable (flux) at the boundaries of the problem domain. Physical boundaries of ground water flow systems are formed by an impermeable body of rock or a large body of surface water. Hydraulic boundaries such as groundwater divides or no flow boundaries represented by chosen stream lines can be used as model boundary. However, stress can change the flow pattern and shift the position of stream lines; therefore care must be taken when using streamline as the outer boundary of a model. In physical terms, for groundwater flow applications, the boundary conditions are generally of three types: (1) specified head, (2) specified flux, and (3) head-dependent flux.

2.6.3 The Advection-Dispersion Equation

The purpose of solute transport model that simulates solute transport in groundwater is to compute the concentration of a dissolved chemical species in an aquifer at any specified place and time. Groundwater velocities calculated using Darcy's Law is used as input to contaminant transport models. In most cases the flow and contaminant transport equations are uncoupled. Due to the assumption that changes in concentration do not affect the flow field, simultaneous solution is not required. The only exception is the case of density dependent flow such as seawater intrusion where changes in concentration affects the density of the water and hence the velocity distribution. The advection-dispersion contaminant transport in three dimensions with internal sources or sinks can be written as Equation 2.2.

$$\frac{\partial C}{\partial t} = \frac{\partial}{\partial xi}(D_{ij}\frac{\partial C}{\partial xj}) - \frac{\partial}{\partial xi}(v_i C) + \frac{q_s}{\theta} C_s \qquad (2.2)$$

Where:

D_{ij} is the dispersion coefficient tensor, v is the mean groundwater velocity or seepage in the x, y and z direction (Darcy velocity divided by effective porosity), C is concentration (mass per unit volume of solution), θ is effective porosity, C_s is source or sink concentration (mass per unit volume of solution), and q_s is the flow rate per unit volume of the source or sink, x is the spatial coordinate, and t is the time. This equation is an expression of the mass balance of a contaminant within the aquifer as a result of change in storage, dispersion, advection and source or sinks. These processes are represented respectively by the first, second, third and fourth terms of the above equation.

2.6.4 Contaminant Transport with chemical reactions

The effects of chemical reactions on solute transport are generally incorporated in the advection-dispersion through a chemical sink/source term. A chemical sink/source term can be formulated for each chemical species or component of interest, and added to the general advection-dispersion equation and written as Equation 2.3 (Zheng and Bennett, 1995).

$$\frac{\partial C}{\partial t} = \frac{\partial}{\partial xi}(D_{ij}\frac{\partial C}{\partial xj}) - \frac{\partial}{\partial xi}(v_i C) + \frac{q_s}{\theta}C + \sum_{k=1}^{N} R_k \qquad (2.3)$$

Where: R_k is the chemical sink/source term representing the rate of change in solute mass of a particular species due to N chemical reactions, and C is understood as the concentration of that species.

The two types of reaction which have frequently been incorporated in advection-dispersion transport models are: equilibrium-controlled sorption reactions, which involve the transfer of mass between the dissolved phases and the solid matrix of the porous medium under the local equilibrium assumption; and chemical reactions, such as radioactive decay and some forms of biodegradation, which can be described by an irreversible first-order rate law. The governing equation that describe three-dimensional transport of a single chemical constitutes in groundwater considering advection, dispersion, fluid sinks/sources, equilibrium-controlled first order irreversible rate reactions written as Equation 2.4.

$$\frac{\partial C}{\partial t} = \frac{\partial}{\partial xi}(D_{ij}\frac{\partial C}{\partial xj}) - \frac{\partial}{\partial xi}(v_i C) + \frac{q_s}{\theta}C_s - \lambda(C + \frac{\rho_b}{\theta}\bar{C}) \tag{2.4}$$

Where:

C is the dissolved concentration, $\bar{C}$ is the sorbed concentration a function of the dissolved concentration, C, as defined by a sorption isotherm, v_i is the seepage velocity, Dij is the dispersion coefficient tensor, Qs is the flow rate of a fluid source or sink per unit aquifer volume, Cs is the concentration of the fluid source or sink flux, λ is the reaction rate constant, θ is the porosity, ρ_b is the bulk density of the porous medium.

2.6.5   Numerical Solutions of the Advection-Dispersion Equation

The advection-dispersion equations can be solved by either analytical or numerically methods. Analytical methods involve the solution of the partial differential equations using calculus, based on the initial and boundary value conditions. They are limited to simple geometry and in general require that the aquifer be homogeneous. Analytical solutions are useful in that they can be solved easily. Numerical solutions are more powerful than analytical solutions in the sense that aquifers of any geometry can be analyzed and aquifer heterogeneities can be accommodated and allow complicated boundary and initial conditions to be included.

Numerical methods are based on a discretisation of the spatial and temporal solution domain, and subsequent calculation of the concentration at discrete nodes in the domain. Consistency, stability, convergence, boundedness and conservation are the most important properties of numerical solution methods (Ferziger and Perić, 1996). The

27

discretisation method for the simulation must be consistent (with the analytical problem), which means that the local truncation error should tend to zero as the grid spacing tends to zero. Numerical method must be stable; it should not magnify the error in the course of the numerical solution process. Additionally, the method must be convergent i.e., the solution resulting from discretisation method has to converge towards the solution of the differential equation as the discretisation steps sizes in space and time tend to zero. Numerical solutions should also lie within proper bounds. Physically non-negative quantities (like concentration) must always be positive. If the boundedness is violated, this can cause an over-or undershooting of the solution (non-monotonic solution behaviour). Since the equations to be solved are conservation laws, the numerical schemes should also-on both a local and a global basis - respect these laws. The method is conservative if the total source or sink in the domain is equal to the net flux of the conserved quantity through the boundaries. This is an important property of the solution method, since it imposes a constraint on the solution error. Non-conservative schemes can produce artificial sources and sinks changing the balance both locally and globally.

The solute-transport equation is in general more difficult to solve numerically than the groundwater flow equation, largely because certain mathematical properties of the transport equation vary depending upon which terms in the equation are dominant in a particular situation (Konikow and Mercer, 1988). According to Konikow and Mercer (1988) and Konikow (2011) when solute transport model is dominated by advective transport, as is common in many field problem, the advection-dispersion equation approximates a hyperbolic type of equation, similar to equations describing the propagation of a wave or of a shock front. On the other hand, if the system is dominated by dispersive fluxes, case where fluid velocities are relatively low and aquifer dispersivity are relatively high, the advection-dispersion equation become more parabolic in nature (similar to the groundwater flow equation). The fact that groundwater flow velocity is varying from zero in low permeability zone to several meters per day in high permeability areas or near stress points makes the transport process to vary between the hyperbolic and parabolic type of equations and make it difficult to solve the entire domain of a single problem to be solved by single numerical method.

When the transport equation is advection-dominated, solutions of the advection-dispersion equation by classical methods (FDM, FEM and FVM) introduce excessive numerical dispersion and oscillations. Numerical dispersion results in an artificial increase in physical dispersion or smearing of sharp concentration fronts. On the other hand oscillations results in overshoot and undershoot about the true concentration. These problems are eliminated by using refined grid in space and time. However, this is not always practical, because of considerable increase in computational time. Two alternative numerical methods have been developed to solve the advection-dispersion problem without numerical dispersion and oscillation problems (Siegel et al., 1997). The first method is called the characteristics method or the Eulerian-Lagrangian method, and the second approach is known as Eulerian approach. In the first approach the advection is decoupled from the dispersion term in the transport equation (Neuman, 1981). Method of characteristics and modified method of characteristics are examples. In the second approaches special schemes are developed to solve the hyperbolic equation with FDM, FEM and FVM methods (e.g., upwind schemes in FDM or FVM). Putti et al.(1990) used FVM approaches based on triangular element and high resolution upwind schemes for solving the general transport equation in porous media. Total variation diminishing schemes (TVD) have been developed by different researchers for solving the advection-dispersion equations (Cox and Nishikawa, 1991; Hou et al., 2012). TVD schemes guarantee that the total variation of a quantity C not to increase as the solution progresses in time (i.e. TV (C^{n+1}) <= TV (C^n)) (Cox and Nishikawa, 1991) This property prevents the formation of oscillation in the solution.

Eulerian-Lagrangian approaches are based on the principle of combining the advantage of simplicity of fixed Eulerian grid (FDM, FEM and FVM) with Lagrangian methods (method of characteristics) which are more suited to advection dominated problems (Konikow and Mercer, 1988; Neuman, 1981). Many numerical methods have been developed in the last decades to solve advection-dispersion problems based on this approaches. Douglas and Russell (1982) presented detailed discussion of numerical methods for advection dominated dispersion problems based on combined method of method of characteristics, FDM and FEM. Neuman (1984) developed an adaptive Eulerian-Lagrangian finite element method based on Neuman (1981) approach for solving the advection-dispersion problems. The idea behind this approach is automatic adaptation of the solution process to the nature of the problem. In areas where sharp concentration present the advective component is tracked forward with the aid of moving particles clustered around each front and away from such fronts, the advection problem is handled by an efficient modified method of characteristics called single-step reverse particle tracking method. When a front dissipates as a result of dispersion, sorption or decay, its forward tracking stops automatically and the corresponding particles are eliminated. The two dimensional test dispersion problem in a uniform velocity field demonstrated that the adaptive method is capable of handling wider range of Peclet number from zero to infinity and large time steps with Courant numbers much higher than 1. Celia et al (1990) develop an Eulerian-Lagrangian Localized Adjoint Method (ELLAM) for the advection-dispersion equation. The approach is based on a space-time discretisation in which specialized test functions which locally satisfy the homogenous adjoint equation within each element. Healy and Russell (1993) developed a Finite-volume Eluerian-Lagrangian method for solving the one dimensional advection-dispersion equation based on Celia et al (1990). The conclusion of their work is that their method is efficient in solving strongly advection dominated problems and for dispersion dominated problems their method gave comparable results with the standard centred FDM, but slightly less efficient in terms of computation time. In addition they reported that their method is mass conservative and its application is less restricted by the size of the Courant number. It is beyond the scope of this study to provide full coverage of these techniques. Other discretisation methods which maintain the monoticity of the solution such as slope-limiter, Flux-limiter methods are briefly described in Hinkelmann (2005).

There are additional complications when the solutes of interest are reactive. The reaction terms included in advection-dispersion equation are mathematically simple ones. They do not necessarily represent accurately the true complexities of the reactions. Also, particularly difficult problems arise when reaction terms are highly nonlinear, or if the concentration of the solute of interest is strongly dependent on the concentration of numerous other chemical constitutes.

2.6.6   RT3D Model

RT3D is a three-dimensional Multispecies Reactive Transport model developed at the Pacific Northwest National Laboratories (Clement, 1997). RT3D is a three dimensional finite difference code that models the advection, dispersion, and reaction or biodegradation of contaminants in the subsurface. RT3D was developed to serve as a modelling tool for sites implementing intrinsic bioremediation, but it can also be used to evaluate contaminant fate and transport under many other reactive or non reactive conditions. RT3D can also accommodate multiple sorbed and aqueous-phase species with any reaction framework that the user needs to define. The current version of RT3D has several pre-programmed reaction modules which include instantaneous aerobic decay of benzene, toluene, ethyl benzene, and xylene (BTEX), degradation of BTEX using electron acceptors, rate limited sorption reactions, a double Monod kinetic model, and sequential degradation of perchloroethylene. RT3D also has user-defined reaction option that can be used to simulate any other types of user-specified reactive transport systems. This allows, for example, natural attenuation processes or an active remediation to be evaluated (Clement et al., 2000). Several numerical codes are available for modelling field scale transport in porous media. However, only RT3D can simulate three dimensional multi-species reactive transports of CAHs.

Governing Equations

The governing equation that describes the fate and transport of aqueous and solid phase species in three dimensional saturated porous media is written as (Clement, 1997; Clement et al., 2000):

$$\frac{\partial C_k}{\partial t} = \frac{\partial}{\partial x_i}\left(D_{ij}\frac{\partial C_k}{\partial x_j}\right) - \frac{\partial}{\partial x_i}(vi\, C_k) + \frac{qs}{n}C_{s_k} + r_c \text{ where } k = 1,2 \dots m \tag{2.5}$$

$$\frac{\partial \tilde{C}_{im}}{\partial t} = \tilde{r}_c, \text{where } im = 1, 2 \tag{2.6}$$

Where:

n is the total number of species, m is the total number of aqueous phase (mobile) species (thus, n minus m is the total number of solid phase or immobile species), C_k is the aqueous phase concentration of the kth species, $\tilde{C}_{im}$ is the solid-phase concentration of the im^{th} species (contaminant mass per unit mass of porous media or contaminant mass per unit aqueous phase volume), D_{ij} is the hydrodynamic dispersion coefficient, v is the pore velocity, n is the soil porosity, qs is the volumetric flux of water per unit volume of aquifer representing sources and sinks, C_s is the concentration of sources/sinks, r_c represents the rate of all reactions that occur in the aqueous phase and $\tilde{r}_c$ represents the rate of all reactions that occur in the solid phase.

Numerical Solutions Procedure

RT3D code uses a reaction Operator-Split (OS) numerical strategy to solve any number of transport equations (in the form 2.5 and 2.6). The OS approach is a way of splitting the reaction terms. This involves dividing the governing equation, the mobile species transport into four distinct equations:

The advection equation
$$\frac{\partial C}{\partial t} = -\frac{\partial}{\partial x_i}(vi\, C_k) \tag{2.7}$$

The dispersion equation
$$\frac{\partial C}{\partial t} = \frac{\partial}{\partial x_i}\left(D_{ij}\frac{\partial C_k}{\partial x_j}\right) \tag{2.8}$$

The source/sink-mixing equation
$$\frac{\partial C}{\partial t} = \frac{qs}{n}C_s \tag{2.9}$$

And the reaction equation

31

$$\frac{\partial c}{\partial t} = r \tag{2.10}$$

Where: the term r represents all possible reaction terms that appears in a typical mobile species transport equation. In a typical immobile species equation (in the form 2.6), the advection, dispersion, and source sink mixing terms are zero, and only the reaction term exists.

RT3D applies this OS strategy using transport routine from US EPA code MT3D to solve the advection, dispersion, and source/sink-mixing equations, and then solves the reactions (Clement *et al.*, 1998). The MT3D code can be used only to describe single-species transport with first order reaction (Clement et al., 1998; Zheng, 1990), but with RT3D multi-species reactive transport can be simulated. The OS solution strategy introduces an error since the coupling of the two sets is neglected, but it can be minimized by choosing small time steps (Schreuder, 2009). The numerical solution procedure follows a sequential step, where the advection, dispersion, and source sink mixing steps for all mobile components within a transport time step are sequentially solved and the reactions of the mobile and immobile phases are solved together. In order to integrate the reaction time steps within transport time step RT3D uses an automated LSODEA differential equation solver (Clement et al., 1998).

2.7   Groundwater-Surface water Interactions

Groundwater and surface water are in a continuous dynamic interaction in the hydrological cycle (Sophocleous, 2002; Winter et al., 1998). These dynamic interactions have practical consequences both for the quality and quantity of water in either system. Traditionally, hydrologists treated stream and groundwater as distinct, independent resources. However, with increasing development of land and water resources, it became apparent that the development of either of these resources affects the quantity and quality of both systems (Winter et al., 1998). For instance, contaminated aquifers that discharge into a stream may cause long-term contamination of the surface water, or, on the other hand, surface water bodies may be a major source of pollution to aquifers. Nowadays, understanding the connection between surface water and groundwater has received growing interest from the research community. The new European Union Water Framework Directive (2000), has recognised the importance of groundwater-surface water interactions, and the need for integrated management of surface water and groundwater bodies. According to Winter et al. (1998), streams connected with groundwater interact in three basic ways: (1) water may flow from groundwater to streams through the streambed (gaining streams) (Figure 2.2a), (2) streams may lose water to the groundwater system (losing streams) (Figure 2.2b),, (3) streams may be gaining in some reaches and losing in other reaches. Streams can also be separated from the groundwater system by unsaturated zone which is known as disconnected streams (Figure 2.2c). Rapid rise in stream stage due to storm precipitation, rapid snowmelt or release of water from the reservoir may causes water to move from the stream into the streambanks. This process is known as bank storage and shown in Figure 2.2d.

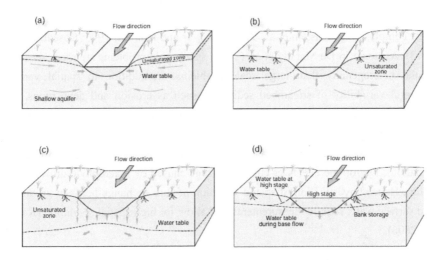

Figure 2.2: Schematic representation of a gaining river (a), a losing river (b), disconnected river (c), and bank storage (d) (Winter et al., 1998).

2.8 Spatial and Temporal Variability of Groundwater-Surface water Interactions

Traditionally, the focus of studies including river-aquifer interaction was on a regional or watershed scale (Fleckenstein et al., 2006). In this context the main interest is to quantify the contribution of the river-aquifer exchange to the overall water balance of the basin. Although, this approach is sufficient for regional scale water management, it is not applicable to investigate the ecological dynamics of river-aquifer systems (Woessner, 2000). More recently numerous researchers have focused on the local scale river aquifer interaction and the role of hypoheric zone in stream ecosystems (Hancock, 2002; Hayashi and Rosenberry, 2002; Storey et al., 2003; Woessner, 2000). Interest towards investigating the local scale groundwater surface water interaction is growing due to many reasons including (Schmidt, 2009): (i) for indentifying the dominant groundwater discharge zones along the river reach that control flow and contaminant transport; (ii) to understand the biogeochemical conditions in the streambed, and (iii) to characterize the groundwater dependent benthic and a aquatic life in the hyporheic zone.

At the watershed scale, groundwater flow patterns and their interaction with surface water are influenced by topography, geology and climate (Tóth, 1970). On the other hand local scale groundwater-surface water interaction is mainly controlled by hydraulic gradient, and hydraulic conductivity of the aquifer and streambed (Woessner, 2000). Owing to the variations in hydraulic gradient and in hydraulic conductivity (caused by heterogeneity), quantifying flow exchange between groundwater and surface water is quite complex (Fleckenstein et al., 2010; Jones Jr and Holmes, 1996; Rosenberry et al., 2013). As noted in De Marsily et al. (2005) small scale problems are more sensitive to spatial heterogeneity than large scale problems. In addition to spatial variability temporal variability of streambed sediments such as erosion and depositions, pools, riffles, streambed topography also affect groundwater surface water interaction at a local scale (Harvey and Bencala, 1993; Rehg et al., 2005). Temporal variability of fluxes may occur as a result of change in hydraulic gradient due to groundwater head variations, hydrologic events or change in stage caused by macrophyte growth (Green, 2006). Hence, investigation of spatial variation of flux exchange between groundwater and surface water require measurements that allow for high spatial resolution (Kalbus et al., 2006).

2.9   Methods for quantifying Groundwater-Surface water Interactions

Extensive review of the different methods available for assessing groundwater surface water interactions is provided by Kalbus et al. (2006). The authors summarized the different available methods in four categories namely: the direct estimation method, the mass balance approaches, Darcy's Law, and heat tracer method. Direct estimation method involves the direct measurement of water flux across the groundwater-surface water interface by using seepage meters. The mass balance approach for groundwater surface water interactions is further classified as incremental streamflow method, hydrograph separation method and environmental tracer method (Kalbus et al., 2006). The incremental streamflow method is based on the concept of the difference in streamflow measured at two successive cross-sections. Hydrograph separation method is the most widely used standard method in hydrology for estimating the base flow component of the hydrograph. The method involves separating the hydrograph measured at a gauging station in different hydrograph components such as base flow and quick flow. Base flow is then assumed to represent the groundwater discharge.

Environmental tracers such as stable isotope and geochemical tracers are in wide use for stream flow origin determination, groundwater dating and separating hydrograph components (Kendall and MacDonnell, 1998).

Groundwater-surface water fluxes exchange estimation using Darcy's Law involves determination of hydraulic gradient and hydraulic conductivity. Darcy flux also known as Darcy velocity is computed by multiplying hydraulic gradient with hydraulic conductivity. Darcy Law is the standard way of formulating groundwater-surface water fluxes exchange mechanisms in most standard groundwater models. The actual water velocity between any two points is computed by dividing the Darcy velocity by porosity. Heat tracer has long been used for quantifying groundwater surface water flux exchange. Darcy's Law based on MODFLOW model and heat tracer methods are further described in subsequent sections.

2.10 River-Aquifer interaction using MODFLOW

MODFLOW is a widely used, finite-difference flow model for simulating saturated flow and is capable of simulating the interaction between rivers and underlying alluvial aquifers. MODFLOW models river-aquifer interaction using RIVER or STREAM packages. The STREAM Package is a streamflow routing model limited to steady flow through a rectangular, prismatic channel (Prudic, 1989). In MODFLOW the RIVER package is the widely used method to quantify flux between the river and aquifer. It assumes that the stream stage remains constant throughout a given stress period within the model. This constant stream stages is then utilized to calculate the flux between the river and aquifer. The head along the stream is determined from the depth of flow in the channel. Streams are divided in reaches and segments, where a reach is a section of a stream associated with a particular finite difference cell. A particular finite-difference cell can have more than one stream reach, but only one finite-difference cell can be assigned to a single reach. River-aquifer exchange is then simulated between each reach and the model cell that contains that reach. Under this condition MODFLOW computes the flow rate between a river and aquifer from the following relationship (i.e. flow between a river reach n and a node i, j, k).

$$QRVI_n = CRIV_n\left(HRIV_n - h_{i,j,k}\right), \qquad h_{i,j,k} > RBot_n \qquad (2.11)$$

$$QRVI_n = CRIV_n\left(HRIV_n - RBot_n\right), \qquad h_{i,j,k} \le RBot_n) \qquad (2.12)$$

Where:

$QRVI_n$ is the flow between the river and aquifer, taken as positive if it directed into the aquifer; $HRIV_n$ is the water level (stage) in the river; $CRIV_n$ is the hydraulic conductance of the river-aquifer interconnection; $RBot_n$ is the river bed bottom elevation; and $h_{i,j,k}$ is the head at the node in the cell underlying the river reach.

$$CRIV_n = \frac{K_n L_n W_n}{M_n}$$

(2.13)

Where: L_n is the length of the river reach as it crosses the cell; W_n is the river width; M_n is the thickness of the river bed sediments; K_n is the vertical hydraulic conductivity of the river bed material.

2.11 Heat as a Tracer

Stream temperature may vary strongly on a daily and seasonal basis because of solar-driven temperature fluctuations at the land surface, whereas groundwater temperature is relatively stable throughout the year. This difference in temperature between streams and groundwater provides a means for tracing exchanges between the two systems. The key processes that are considered in such studies are advection of heat with the flowing water and conduction of heat through the solid-water medium (Stonestrom and Constantz, 2003). The concept of using heat as a natural tracer goes back to the early 1900s. Since then, the theory has been expanded and numerous methodologies ranging from simple analytical methods to complex numerical models have been developed. A comprehensive review of heat as a natural tracer of groundwater can be found in Anderson (2005). Suzuki (1960) was among the earlier researchers to recognize the use of temperature as a natural tracer. Following the work of Suzuki (1960), Stallman (1965) developed a modified analytical solution for computing infiltration rates from the transient temperatures observed near the land surface. Bredehoeft and Papaopulos (1965) also developed an analytical solution for describing vertical steady flow of groundwater and heat through an isotropic, homogenous, and fully saturated semi-confining layer.

Many authors, e.g. Healy and Ronan (1996), Kipp (1987), Voss (1984), have expanded the theory of heat flow and developed numerical models to solve the unsteady heat flow condition. These numerical models are advantageous in the sense that they can represent more complicated boundary conditions in multiple dimensions, incorporate heterogeneity in sediment properties and can accommodate complex geometries. Generic codes are available to solve the coupled groundwater flow and heat transport problems. These codes can be applied to a variety of unsaturated and saturated zone flow and transport studies such as groundwater recharge, exchange of surface and groundwater, and transport of spilled contaminants (e.g. VS2DI, Healy (2008)).

Temperature measurements have been used as a natural tracer in a wider area of application in groundwater studies such as groundwater-surface water exchange studies ranging from perennial streams (Lapham, 1989; Stonestrom and Constantz, 2003) to ephemeral channels (Ronan et al., 1998), calculating groundwater flux to a lake (Anibas et al., 2009), identifying spatial variability of groundwater discharge in a wetland stream (Lowry et al., 2007), studies of surface warming effect on groundwater conditions (Ferguson and Woodbury, 2005; Taniguchi, 2006; Taniguchi et al., 1999; Taniguchi and Uemura, 2005), basin scale studies through groundwater temperature mapping for identifying groundwater recharge zones (Cartwright, 1970; Schmidt et al., 2007), and for delineating hyporheic zone flow (the region of mixing between groundwater and surface water (Conant Jr, 2004). Hyporheic zones play a critical role in governing contaminant exchange and transformation during water exchange, and hence are very important for ecological impact studies (Alexander and Caissie, 2003).

2.12 VS2DH Model

VS2DH is a finite- difference variably saturated two-dimensional ground-water flow and heat transport model developed by Healy and Ronana (1990). It solves Richards' equation for water flow (Richards, 1931) and the advection-dispersion equation for heat transport. The advection-dispersion equation is formulated in terms of temperature and is calculated using pore water velocity computed from Richards' equation considering the temperature effect on hydraulic conductivity. For more detailed descriptions see Healy (1990), Healy and Ronan (1996) and Kipp (1987).

Governing Equations
Heat transport through variably saturated material is described by the advection-dispersion equation (Equation 2.14). This equation is derived by balancing the changes in energy stored within a volume of porous media due to ambient water of different temperature flowing into the volume, thermal conduction into or out of the volume, and energy dispersion into or out of the volume (Healy and Ronan, 1996).

$$\frac{\partial[\theta C_w + (1-\Phi)C_s]T}{\partial t} = \nabla.K_T(\theta)\nabla T + \nabla.\theta C_w D_H \nabla T - \nabla.\theta C_w v T + q C_w T^* \qquad (2.14)$$

Where: t is time; θ is volumetric moisture content; C_w is heat capacity of water; Φ is porosity; C_s is heat capacity of dry solid; T is temperature; K_T is the thermal conductivity of the water and solid matrix (a tensor); D_H is the hydrodynamic dispersion tensor; v is water velocity; q is rate of fluid source; and T^* is temperature of fluid source.

The left-hand side of Equation (2.14) represents the change in energy stored in a volume over time. The first term on the right-hand side of the equation reflects energy transport by thermal conduction. The second term on the right side represents transport due to thermo-mechanical dispersion. The third term on the right side accounts for advective transport of energy. The last term represents heat sources or sinks. The thermo mechanical dispersion is governed by the dispersion coefficient D_H which can be related to the groundwater velocity with the Equation 2.15 (Healy, 1990).

$$D_{Hi,j} = \alpha_T |v| \delta_{ij} + \frac{(\alpha_L - \alpha_T) v_i v_j}{|v|} \tag{2.15}$$

Where $D_{Hi,j}$ is the i, j-th component of D_H; α_T is the transverse dispersivity; α_L is the longitudinal dispersivity; δ_{ij} is the Kroneker delta function (equal to 1 when i=j and equal to zero otherwise); v_i is the i_{th} component of the velocity vector and v_j is the j_{th} component of the velocity vector.

The flow velocity within variably saturated sediments for Equation 2.14 is determined by solving Richards' equation (Equation 2.16), identical to the one given by Lappala et al.(1987). In this equation total hydraulic potential is used as the dependent variable so as to allow straightforward treatment of both saturated and unsaturated conditions.

$$v\{\rho[c_m + sS_s]\}\frac{\partial H}{\partial t} - \rho \sum_{k=1}^{\widehat{m}} A_K \, KK_r(h)\frac{\partial H}{\partial n_K} - \rho q v = 0 \tag{2.16}$$

Where: v is the volume of porous medium; ρ is the liquid density; c_m is the specific moisture capacity; s is the liquid saturation; S_S is specific storage; t is time; q is volumetric source-sink term; A_K is the area of the k_{th} face to which n_K is orthogonal; $K_r(h)$ is relative hydraulic conductivity as a function of pressure potential (dimensionless); K is saturated hydraulic conductivity; H is the total hydraulic potential and $\widehat{m}$ is the faces of a general curvilinear polygonal volume.

Equation 2.16 states that the rate of change of mass stored in a given volume of porous medium v is balanced by the sum of liquid flux across the surface boundary of v and liquid added or removed by sources and sinks. Healy and Ronan (1996) implemented the same equation in VS2DH but by taking into account the temperature dependence of saturated hydraulic conductivity. Because of the temperature dependence of viscosity they have treated saturated hydraulic conductivity K as a function of temperature (Equation 2.17).

$$K = \frac{\rho g k}{\mu(T)} \tag{2.17}$$

Where: ρ is density, g is gravity, k is intrinsic permeability, and μ is viscosity.
Viscosity is calculated according to the empirical formula Kipp (1987) (Equation 2.18).

$$\mu(T) = 0.00002414 * 10^{247.8/(T+133.16)} \tag{2.18}$$

Because of the small variation in density across the range of subsurface temperatures in shallow sediments the density of water is typically treated as constant in VS2DH.

The hydraulic conductivity can be written as (Equation 2.19)
$$k(T, \psi, x) = ksat(T, x) * kr(\psi, x) \tag{2.19}$$

Where T is time; x is the vector of spatial dimension; ψ is the pressure head; $k(\psi, x)$ is the hydraulic conductivity; $ksat(T, x)$ is the temperature dependent hydraulic conductivity and $kr(\psi, x)$ is a relative hydraulic conductivity; the ratio of unsaturated to saturated hydraulic conductivity.

2.13 Mathematical Optimisation

Formulation of optimisation problem needs definition of management objectives, decision variables, and management constraints. Management objective also known as an objective function is described by an equation, whose value is minimized or maximized during optimisation to achieve optimal values. Decision variables are variables, whose values will be determined as part of the optimisation solution. State variables describe physical response to change in stress. Constraints are limits on acceptable values of decision and state variables.

Depending on the types of mathematical relationships between the objective functions, constrains, and decision variables the solution methods or algorithms used to solve the optimisation problem are different. The common optimisation problem types include: linear programming, quadratic programming, mixed-integer programming and nonlinear programming problems. Linear programming problems are ones in which the objective function and all the constraints are linear function of the decision variables. LP problems are easier to solve than other types of optimisation problems. Quadratic programming (QP) problems have objective functions which are a quadratic function of the decision variables, and constraints which are linear function of the variables. Mixed-integer (MIP) problems are those optimisation problems where the decision variables are constrained to be an integer values at the optimal solution. A special case of decision variables used in MIP problems are either 0 or 1 solution. Such variables are called binary integer variables and can be used to model yes/no decision, (such as, for example whether to turn on or off a pumping well). However, integer variables make the optimisation problem nonlinear problem which is difficult to solve and require more memory.

Nonlinear optimisation (NLP) problems are those optimisation problems that involve the objective function or at least one of the constraints are nonlinear function of the decision variables. QP problems are special case of NLP problems. NLP problems are non-convex optimisation problem types where multiple feasible regions and multiple locally optimal points within such region exist. Global optimisation (GO) methods are designed to solve such type of problems. They do not require the function to be differentiable or the variables to be continuous (or even to be explicitly known, e.g when the function values are provided by a simulation model), which are the case in the gradient based solvers. Gradient based solvers compute the gradient of the function at various trail solutions, and move in the direction of the negative gradient when minimizing; and move in the direction of positive gradient when maximizing. The point where a gradient based algorithms terminates depends on the location where they started; hence they can only find local optimal solution in the vicinity of the starting point. The use of GO avoids becoming trapped at local optimum and finds better solutions.

2.13.1 Global Optimisation Tool GLOBE

GLOBE is a PC-based optimisation tool developed by Solomatine et al. (1999) for finding the minimum of a function of multiple variables. GLOBE was originally used for model calibration problems. Maskey et al. (2002) has extended its application to groundwater remediation problem and solved two groundwater remediation objective functions: minimizing aquifer cleanup time and pumping cost. GLOBE can be easily adapted to be coupled with an external program (simulation model) that supplies objective function value. The second advantage of GLOBE is that it has many global optimisation algorithms that the user can choose and compare the results of those different algorithms. The current version of GLOBE consists of nine global optimisation algorithms, which include: two controlled random search algorithms (CRS2 and CRS4), two multistart algorithms (Multis and M-simplex), adaptive cluster covering (ACCO), adaptive cluster covering with local searches (ACCOL), adaptive cluster covering with decent (ACD), adaptive cluster covering with descent and local searches (ACDL), and Genetic algorithms (GA).

2.13.2 Genetic Algorithms

Genetic Algorithms are class of evolutionary algorithms that are based on the idea of natural evolutions. GA uses a random search procedure inspired by biological evolution, cross-breeding trial designs and allowing only the fittest design to survive and propagate to successive generations (Gen and Cheng, 2000). Objective function values are used to determine the candidate solution's relative fitness and to guide GA towards the evolution of good solutions. GA uses the following steps to evolve solutions to the search problem (Sastry et al., 2005): i) Initialization; the initial population of candidate solutions is usually generated randomly across the search space; ii) Evaluation; the fitness values of the candidate solutions are evaluated; iii) Selection; based on the fitness values selection is performed in which the most fit members of the population survive and the least fit members are eliminated; iv) crossover; pair of parent individuals selected based on their fitness or function values are recombined to generate a new offspring. v) Mutation; while crossover operates on two or more parental chromosomes, mutation randomly change their position in space locally; vi) Replacement; the offspring population created by selection, crossover, and mutation replaces the original parental population become the next generation. Steps ii-vi is repeated until a terminating condition is met.

As described above genetic algorithm is a population based algorithm that uses selection, crossover and mutation operators to generate new sample points in search space. Hence its performance largely depends on population size, number of generations, and the probabilities of crossover and mutation (McKinney and Lin, 1994). For instance, while small population sizes lead to premature convergence, large population size on the other hand lead to unnecessary computational time. McKinney and Lin (1994) suggested that population size of 50-100 and generation of 10-20 is enough to get optimal or near optimal solution. Detail description about genetic algorithms can be obtained from Sastry et al. (2005).

2.13.3 Controlled Random Search

The controlled random search (CRS4) algorithm (Ali and Storey 1994) is a modified form of controlled random search (CRS) algorithm proposed by Price (1983). The principle of the original form of CRS algorithm is that the search region (V) is assumed to be defined by specifying upper and lower limits to the domain of each variable. At each iteration, a simplex is formed from a sample, and a new trial point is generated as a reflection of one point in the centroid of the other points in this simplex. If the worst point in the initially generated set is worse than the new one, it is replaced by the latter. In CRS4, two new ideas were incorporated into the original CRS algorithm. The uniform distribution that was used to select the initial sample was modified to Hammersley distribution, and the new points are sampled from the multidimensional beta-distribution.

2.14 Combined Simulation-Optimisation Modelling Approaches

There are two different methods of combining a groundwater simulation model with an optimisation technique: The embedding method and the response matrix method. In the embedding method the numerical discretisations of the partial differential equations are included as constraints in an optimisation algorithm. In this method, the unknown groundwater variables (heads and source/ sink) become decision variables in the optimisation method. Because each element within the modelled domain is represented by an equation in the optimisation problem, the resulting optimisation problem is usually huge, hence the main disadvantages to its application. The response matrix technique utilizes superposition and linear systems theory to simulate groundwater flow (Ahlfeld and Mulligan, 2000; Gorelick, 1983). A response matrix consists of linear influence coefficients that describe the response of hydraulic head to a unit volume of extraction or injection of groundwater. These coefficients are termed response functions. The hydraulic head response matrix is formed by running the simulation model once for each injection or pumping well. This helps to obtain hydraulic heads at each control points as a linear combination of the injection or pumping rates. The response function approach is applicable to linear system, notably confined aquifers. In unconfined aquifers, the changing saturated thickness produces nonlinearity in the groundwater system. It has wider application mainly due much smaller dimensions and solutions for larger problems are possible.

2.15 Multi-objective Optimisation

A multi-objective optimisation problem (also referred to as vector optimisation) is characterized by more than one objectives which are to be optimized simultaneously. In most cases the objectives need to be optimized are conflicting. An improvement in any one objective can only be achieved at the expense of sacrificing the objective value of the other one; meaning there is always a tread-off. In multi-objective optimisation, one solution which is best with respect to all objectives is not necessary optimal for other objective function, rather there exists a set of solutions which are superior to the rest of solutions in the search space when all objectives are considered but are inferior to other solutions in the space in one or more objectives. These solutions are known as Pareto-optimal solutions or nondominated solutions (Srinivas and Deb, 1994). Dominance is an important concept in the multi-objective optimisation and can be introduced as follows: A vector u= [u1 u2 u3 ...uk] is said to be dominated by the vector v=[v1,v2, v3...vk] if and only if v is less than u in all the component of the vector, meaning v1 < u1, v2 < u2...vn < un. When at least one component in the vector u is not less than the corresponding in vector v, the vector v does not dominate u. A special dominance concept is weak dominance, and it occurs when some components in the vectors component are equal.

There are several GA based multi-objective optimisation algorithms, and the Non-dominated Sorting Genetic Algorithm II (NSGA-II) is one of the most successfully applied multi-objective optimisation algorithm in many fields. NSGA-II was developed by Deb et al. (2002) as improvement of its antecedent NSGA developed by the same author. In NSGA-II the offspring population is first created by using the parent population, based on mating and mutation processes and these two populations are combined forming a new population of 2N dimension. Then a non-dominated sorting is used to classify the entire population; this classification is made using a ranking (fitness) function to identify several fronts.

2.16 Conclusion

This chapter provided theoretical background required for understanding contaminant transport modelling, groundwater-surface water interactions and mathematical optimisation methods that are necessary for consecutive chapters. It reviewed the general forms of the governing groundwater flow equation, the advection-dispersion equation, governing groundwater flow and heat transport equations. It also described the different types of models used in this study.

3   Modelling Natural Attenuation of Chlorinated Solvents

3.1 Introduction

Chlorinated solvents were used in a wide variety of 20th century industries. The most significant users of these chemicals were aerospace, military, metal-working, and dry-cleaning industries. Because of their low flammability and reactivity, their ability to quickly and efficiently dissolve wide ranges of organic substance made them useful to most industries (Doherty, 2000). The use of chlorinated solvent during the first 60 years of the 20th century was mainly influenced by economic conditions and wartime demand. Increased demand during wartime years was generally followed by a period of post-war oversupply (Morrison et al., 2010). According to Morrison et al. (2010) in the later 40 years, environmental regulations in the USA such as the 1977 Clean Water act, the 1980 Resource Conservation and Recovery Act, and the 1990 Clean Air Act played a great role in amending the use and handling of these compounds.

Depending on their use and origin, the potential releases of these compounds to the environment differs over the course of time. For example, by the late 1940s, tetrachloroethene (PCE) was the most predominant non-petroleum dry-cleaning solvent. The peak years for PCE production in the United States ranged from the late 1960s to the early 1980s (Doherty, 2000). By the year 1980s the effect of environmental regulations and improvements in the dry-cleaning process caused an overall decrease in PCE demand. In many countries, this decrease in demand resulted in an end of PCE manufacturing. In the year between 1930s and 1970s trichloroethene (TCE) was also used as a volatile anesthetic in both Europe and North America. Usage of TCE in dry cleaning decreased in the early 1950s after it was found to degrade cellulose acetate dyes. However, production of TCE has grown until 1970. The decline in production that began in 1970 was the result of increasing evidence of toxicity, economic factors, and increased environmental regulation. Its use as a solvent experienced a rebound in the 1990s when it was listed as a recommended substitute for other solvents (such as TCA) banned under the Montreal Protocol and Clean Air Act Amendments (Kirschner, 1994).

Because of past industrial activities chlorinated solvent contamination has occurred in many parts of the world (National Research Council, 2000; Pankow and Cherry, 1996). Chlorinated solvents such as PCE and TCE, which belongs to chlorinated aliphatic hydrocarbons (CAHs), are common groundwater contaminants (Wood et al., 2006). Contamination of groundwater with PCE or TCE become more problematic due to the fact that a certain groups of anaerobic bacteria transform these compounds to Cis-1,2-

dichloroethylene (cis-DCE) and Vinyl chloride (VC), where the VC is known to be carcinogen to humans (Schaerlaekens et al., 1999). Due to their low solubility in water and less sorption to the soil matrix, these compounds can travel longer distance and contaminant large portion of the aquifer (Pankow and Cherry, 1996). Another unique feature of these compounds is that since, they are denser than water they can travel below the water table, and continue to sink within the aquifer until they reach the impermeable layer.

Due to their potential impact on humans' health there is a considerable interest to remediate CAHs from the aquifer whenever there presence is known. One way to remediate these compounds is to use pump and threat method. But, this method is known to be expensive and in some cases it is not effective (Shoemaker et al., 2003). The other method that got greater acceptance by both industry and decision makers in the last few decades was the use of natural attenuation (National Research Council, 2000). The idea of natural attenuation technique is that natural processes in the subsurface such as biodegradation, dispersion and sorption enable sufficient mass reduction before reaching the down gradient receptor, so that they do not to pose any risk (Ling and Rifai, 2007; Wiedemeier et al., 1999). Natural attenuation is also referred as intrinsic remediation, intrinsic bioremediation, natural recovery or passive remediation (Azadpour Keeley et al., 2001).

Previously published natural attenuation studies such as Clement et al. (2000) and Semprini et al. (1995b) indicated that, in most anaerobic aquifers biological activities that are needed to degrade chlorinated organic compounds are present. The use of natural attenuation as remedy has certain advantages such as: they are cheaper compared to other conventional remedial alternatives, and can be applied in more complex hydrogeological settings without dealing with the uncertainties due to geochemical, bio-chemical, and hydogeological (Azadpour Keeley et al., 2001). One of the main disadvantage is that it may require longer time frame to achieve remediation objectives (National Research Council, 2000; Wiedemeier et al., 1999). The time required to achieve desired remediation objective is also depends on the site conditions.

Groundwater and transport models have been used successfully in the past to support decisions to use natural attenuation processes as the means to achieve remedial objectives at many sites contaminated with CAHs (e.g. Clement et al.(2000), Ling and Rifai (2007)). However, since natural attenuation potential depends on the hydrogeology and geochemistry of a particular site, site specific modelling work is needed to evaluate its success. This part of the thesis is devoted to a natural attenuation modelling case study of chlorinated solvent and their degradation products in the Vilvoorde–Machelen site located in Belgium. In this site, chlorinated solvents primarily consist of CAHs represent the most common organic groundwater contaminant. This study illustrates a modelling approach with sensitivity analysis for of natural attenuation as a sole remedy at the site. This part of the thesis is aimed to address the following research questions.

- How a plume of contamination is evolved, and how it will behave in the future?
- Is the potential for natural attenuation at the site enough for CAHs remediation?
- Is contaminant source zone removal from the saturated zone a good alternative remedial measure to reduce the concentration of CAHs at the site?

3.1.1 Degradation Reactions and pathways

Degradation pathways of chlorinated solvents are important in understanding the fate and transport of these chemicals in the subsurface. The degradation of chlorinated solvents in soil and groundwater occurs by chemical (abiotic) and microbial (biotic) process. The most commonly recognized degradation pathways for chlorinated solvents are those that occur biotically under anaerobic conditions and for less chlorinated compounds (one or two chlorine atoms) under aerobic conditions (Morrison et al., 2010). For example, under anaerobic conditions PCE can be transformed to TCE, to DCE, to VC and to ethene (vogel and McCarty 1985) as shown in Figure 3.1.

According to Nyer (2008), the most important process for the natural biodegradation of the more highly chlorinated solvents is reductive dechlorination. During reductive dechlorination, the chlorinated solvent is used as an electron acceptor, not as a source of carbon, and a chlorine atom is removed and replaced with a hydrogen atom. Reductive dechlorination generally occurs by sequential dechlorination from PCE, to TCE, to DCE, to VC, to ethene. Thus, reductive dechlorination of chlorinated solvents is associated with an accumulation of daughter products and an increase in concentration of chloride ions. Although the three isomers of DCE can theoretically be produced under reductive dechlorination, cis-1, 2-DCE is a more common intermediate and 1,1-dichloroethylene (1,1-DCE) is the least prevalent one (Wiedemeier et al., 1998). According to Wiedemeier et al. (1998) if more than 80% of the total DCE at particular site is cis-1,2-DCE the DCE is likely a reductive dechlorination produced daughter product of TCE.

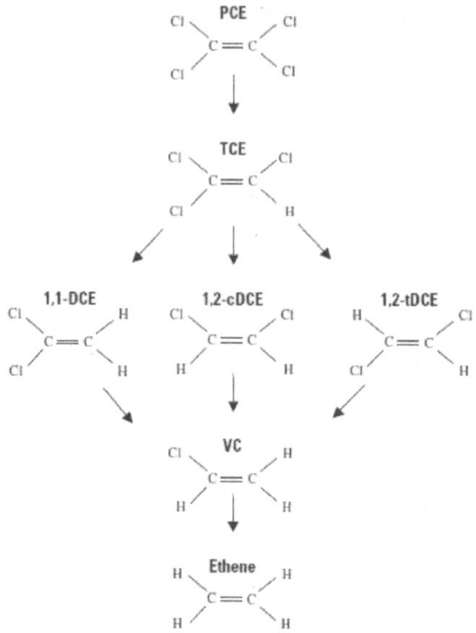

Figure 3.1: Reductive dehalogenation of PCE (adapted from Wiedemeier et al.(1999)).

3.2   Study Area Description

The industrial area of Vilvoorde–Machelen, is located about 10 km North-East of Brussels, Belgium and comprises 10 km^2 in size. It is located in the Zenne catchment along the Zenne River. The Zenne River is 103 km long and has a basin area of 116 km^2. The site lies between the Brussels-Scheldt Maritime Canal and the Zenne River on the west, the R22 road on the east and Trawool River on the north. The topography ranges from 7 to 50 meter above sea level (m.a.s.l), with an average value of 16 m.a.s.l. and mean slope of 1.3 %. The dominant soil type is silty loam, while in the north-eastern part clay-loam exists.

The long industrial history of the Vilvoorde–Machelen area has lead to complex patterns of pollution from multiple sources. The groundwater of the study area is highly contaminated with contaminant from different sources. Bronders et al. (2007) identified four major sources zones of groundwater contaminants shown in Figure 3.2. Source area 1, resulting from the activities of a former paint and varnish factory, was characterized by the presence of a LNAPL (light non-aqueous phase liquid) pool. Source area 2 (a former solvent recycling site), and source area 3 (car manufactory) were characterized by the presence of pure product of both LNAPL, BTEX (benzene, toluene, ethylene, and xylene) and DNAPL (dense non-aqueous phase liquid). Source area 4, located near a paint factory was characterized by only high concentrations of DNAPL. Beside these, it is reported that several other volatile organic hydrocarbon (VOC) plume sources have been identified. However, no clear characterization of the source zones could be given for these plumes. The present study focuses on source area 3. Field investigation in the study area indicated that the groundwater plume is about 70 ha which is roughly 1.2 km x 0.6 km (Bronders et al., 2007). Bronders et al. (2008) also indicated that most of the pollution in the study area roughly occurs from 12 to 14 m of the soil profile. They reported that depth below 14 m is related to less pollution.

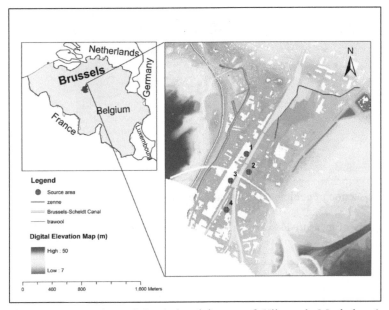

Figure 3.2: Location of the industrial area of Vilvoorde-Machelen (near Brussels), indication of the groundwater plume and the major known source area (modified from Bronders et al. (2007)).

3.3 Geology of the study area

Geologically, the study area is formed by an accumulation of several marine sand deposits (see Figure 3.3 for geological x-sections and locations respectively). The upper layer of the formation is characterized as fine, silty eolian sands (Q1) and the second layer in which the upper layer is based is characterized by the sediments deposit consisting of gravel in a silty sand matrix, with clay–containing organic base (Q2), The wedge-shaped Ghent formation consisting of densely packed sand (GePi) and a local thickness of 4 m, followed by clay-containing silty sand (GeMe) only appears in the north of the study area and the southern edge of this formations roughly coincides with the location of the Woluwe River. The lower layer is characterized by a 25 m thick deposit of silty, glauconite-containing fine sands, known as the Tielt formation

(TiEg).This formation is base on tertiary clay-rich Kortrijk formation. This formation was used as a lower boundary of the hydrogeological model developed in the past because of its low hydraulic conductivity.

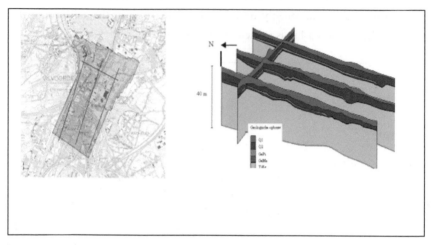

Figure 3.3: Cross section of geological layers adopted from Boel (2008).

Table 3.1: Summary of measured hydraulic conductivity obtained using single well tests (Source Touch et al. 2007).

Well	X	Y	Depth (m) below ground surface	K (m/d)
PB9F1	153599.0	177152.0	6.3	2.2-3.9
PB9F2	153599.5	177152.5	9.3	0.85-1.1
PB16 F1	153808.0	178776.0	7.2	0.08
PB16F2	153808.5	178776.5	11.2	

3.4   Groundwater Flow model

Groundwater flow models are used to calculate the rate and direction of movement of groundwater through the aquifer so that it can be used in a transport model to simulate the movement and chemical alteration of solutes as they move with the groundwater. The groundwater flow system was modelled using the MODFLOW, a widely used modular three-dimensional block-centred finite difference model code (Harbaugh et al., 2000) in Processing MODFLOW modelling environment (Chiang and Kinzelbach, 1998). The conceptual setup of the model, boundary conditions, geologic formations, hydraulic parameters, pumping wells and drainage data were used from previous study (Boel, 2008; Dujardin, 2012; Touchant et al., 2006). A steady state flow field is assumed mainly for two reasons: 1) observed groundwater elevations indicate that the flow system is relatively stable; 2) there are no significant pumping wells conditions within the model domain that influences the groundwater flow enough to cause transient flow conditions.

3.4.1   Model Domain

The model domain covers an area of 10 km^2. The finite-difference mesh consists of 60 rows and 52 columns with a uniform grid size of 50 m x50 m. In the vertical directions, the model consists of five layers with variable layer thickness. The flow model was oriented so that the principal directions of the coordinate axes were aligned with principal direction of regional groundwater flow (i.e. the model grid was rotated at an angle of 30°).

3.4.2   Boundary conditions

The following boundaries were assigned along the perimeter of the active model domain: in the Western side the Brussels–Scheldt navigation canal was specified as general head boundary, in the Southern side, as there are no well-defined hydrologic or physical boundary conditions no flow boundary was assumed, in the Eastern side prescribed head boundary condition was used, and in the Northern side, two boundary conditions were used: Traweol River for some part and no-flow boundaries in the other part. In the vertical direction, the model is bounded by a recharge flux at the top and by impermeable layer at the bottom. The recharge fluxes computed using WetSpa model were obtained from Dujardin et al. (2011).

In the model the Zenne River was represented using the river package of MODFLOW. The river package in MODFLOW needs three parameters: (i) river stage, (ii) river bed elevation, and (iii) conductance of the riverbed material. These values were obtained from Boel (2008).

3.4.3   Model calibration

The groundwater flow model was calibrated using observed head data from 26 monitoring wells measured from 1999 till 2005. The final calibrated head distribution displayed variance of 0.12 m from the observed water table elevations (Figure 3.4). Calibrated water table contours for the second model layer are shown in Figure 3.5. The general orientation of the groundwater flow is towards the north-west direction.

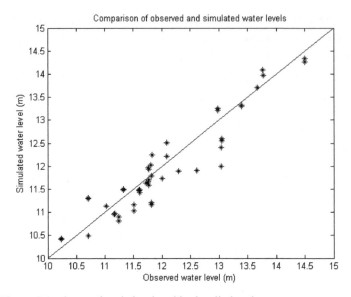

Figure 3.4: observed and simulated hydraulic head.

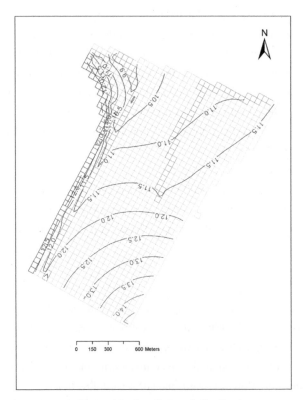

Figure 3.5: Calibrated hydraulic head distribution.

3.5   Reactive Contaminant Transport model

The rate of contaminant transport in groundwater is governed by many factors such as advection, dispersion, retardation and decay. In this study, reactive transport was simulated using the RT3D (Reactive Transport in Three Dimensions) code (Clement, 1997; Clement et al., 2000). The RT3D code used in this study is a part of highly integrated simulation system called Processing MODFLOW. It utilizes the same finite-difference grid as of the MODFLOW model. For numerical stability reasons the model grid close to the pollutant source were refined to 10 m x 10 m so as to eliminate artificial oscillations in the finite difference method (Figure 3.6). The Peclet number, the ratio of velocity multiplied by grid size to the dispersion coefficient, which is a measure of the degree to which the transport problem is dominated by advection is kept smaller than 2.

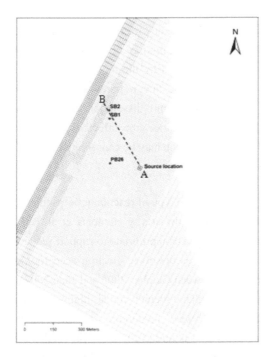

Figure 3.6: Refined grid and source locations.

3.5.1   Model Inputs

Source Information and Initial Conditions

Information related to source is one of the key inputs to any transport model. Similar to other contaminant transport modelling case studies the historical release of contaminant at the site is poorly known. The location of the source, the time period over which the source was active, and the magnitude of the source are all important characteristics which need to be known. Lack of information on contaminant sources may influence the selection and design of a remedial technology or may result costly and lengthy remediation works (Wiedemeier et al., 1996). In order to track the contaminant path and source area we have done backward particle tracking by introducing particles in SB1 and SB2 monitoring wells. The locations of the wells correspond to the source locations are shown in figure 3.6.

PCE and TCE were assumed to be the primary contaminants that were originally disposed at the site. The infiltrations of these substances into the underground were assumed to occur in the period between 1950 and 1990. In Belgium, use of chlorinated solvents was prohibited by law in 1996 (Personal communication). The initial condition was set to zero, assuming clean aquifer before the start of the simulation. PCE and TCE were assumed to be at saturation concentration. That means we assumed pure-phase solubility of PCE and TCE at the source zone. We acknowledge the limitation in our assumption because in most case NAPL may exist as a mixture of chemicals hence the effective solubility (the pure phase solubility multiplied by the mole fraction of each chemical in the NAPL) has to be used. However, this requires accurate NAPL characterization at the source zone. Hence, equilibrium dissolution at pure phase solubility limit was assumed to model the source. Since, the effective solubility is always less than the pure phase solubility our assumption may be conservative.

Reaction Processes and Kinetics

Geochemical data are usually important to identify the type of reaction process that may occur at a particular site. Previous study at the study site by Hamonts et al. (2012) indicated that the geochemical conditions at the site are appropriate to support reductive dechlorination processes. Redox parameters such as dissolved electron acceptors and respiration products measured in groundwater between October 2005 and June 2007 are presented in Table 3.2. The presence of methane demonstrated highly reducing methanogenic conditions in groundwater that were favourable for the observed reductive dechlorination. Reductive dechlorination is one of the principal mechanisms involved in the biodegradation of CAHs under anaerobic conditions (Wiedemeier et al., 1999). In the present study, the sequential dechlorination PCE to TCE, TCE to DCE and DCE to VC is assumed to represent the reaction process in the study site. Of the three possible DCE isomers, the cis isomer of 1,2-dichloroethylene (cis-1,2DCE) predominant in the area, which hereafter is referred as DCE. It could be possible that at the fringes of the plume some aerobic degradation may occur hence we also investigated the Aerobic/Anaerobic reductive dechlorination pathway which was used by Touchant et al. (2006) to model other source in study area. The results (not shown in this thesis) show that the simulated dimension of the plume is more or less the same in both reaction pathways.

Table 3.2: Physicochemical characteristics of groundwater at the Zenne site recorded October 2005 and June 2007 (minimum and maximum values measured for each parameters, d is detection limit) Source: Hamonts et al. (2012).

Observation well and depth	SB2 (7-8 m)	SB2 (9-10 m)	SB1 (7-8 m)	SB1 (9-10 m)
Methane (μg/L)	606-933	638-841	422-809	176-605
SO_4 (mg/L)	208-258	199-247	195-247	120-203
Fe (mg/L)	7.9-9.2	5.6-8.7	5.2-6.4	3.0-4.3
Mn (mg/L)	0.51-0.54	0.36-0.53	0.30-0.39	0.19-0.23
NO_3 (mg/L)	<d	<d	<d	<d
NO_2 (mg/L)	<d	<d	<d	<d
DOC (mg/L)	5-7	2-6	5-8	2-5

Determining the degradation rates is a complex task as they might depend on several biogeochemical and environmental factors such as type and amount of dechlorinating bacteria present in the aquifer and availability of electron donors (Semprini et al., 1995b). According to Macrty (1997) the common electron donors that involve in the reductive dechlorination can be obtained from dissolved organic carbon either from the native soil or other sources such as landfill leachates and petroleum hydrocarbons. For most field scale applications chlorinated solvents degradation reaction are often hypothesized as first-order decay reactions (Clement et al., 2000; Wiedemeier et al., 1996). First-order rate processes are assumed in this study to describe the transformation of these chemical species in the aqueous solution.

Dispersion

The heterogeneities and non-uniformity of the aquifer matrix causes micro-scale spreading commonly referred to as dispersion. Dispersion is a process that describes the spread of contaminants along and perpendicular to the flow direction. Dispersivity at a field site is essential in predicting the transport and spreading of a contaminant plume. In this study we used the guidelines presented in Gelhar et al. (1992) and estimated the field scale dispersivity value for our problem as 10 m. The horizontal and vertical transverse dispersivity was assumed to be 0.1 and 0.01 respectively.

Sorption

Contaminants such as chlorinated solvents can attach (partition or adsorb) to the organic fraction of the aquifer matrix, their transport in the dissolved phase is slowed. In this study we used linear isotherm equilibrium to simulate the sorption. Linear isotherm is more widely used than other types of nonlinear equilibrium isotherms mainly due to its simplicity and convenience of use in practice (Zheng and Bennett, 1995). The linear isotherm uses a single distribution coefficient, kd, to define the relation between the concentration in the dissolved phase and concentration aqueous phase in the porous matrix (kd = mass concentration in the sorbed phase/ mass concentration sorbed in the solid phase). The distribution coefficient is influenced by the fraction of organic matter in the porous media. Site-specific kd values can be calculated multiplying the octanol-water portioning coefficient of the contaminant with the fraction of organic carbon of the soil. The resulting estimates of kd values are then used to compute the retardation coefficient using Equation 3.1. However, the correlation between kd and the aquifer organic content and the contaminant's octanol/water partition coefficient appears to be relatively poor for aquifers with low organic carbon content (i.e. foc < 0.1 %) (McCarty, 1996). For this study representative kd values for each chemical species for each model layers were obtained from Touchant et al. (2006). Similarly, a representative value of effective porosity was determined from the same study to be 20 %. Bulk density of the aquifer material was assumed to be 1.7 kg/l.

$$R = 1 + \frac{\rho_b k_d}{n} \tag{3.1}$$

Where: R is coefficient of retardation, ρ_b is bulk density of aquifer, kd is distribution coefficient and n is porosity.

3.5.2 Model calibration

A transient reactive transport model was calibrated using concentration data from three monitoring wells. Simultaneous calibrations of source strength and degradation rate constants were performed to match the simulated and observed concentration values during the 2005-2007 monitoring effort completed at the study site. The calibration was performed using the automated parameter estimation code PEST (Doherty, 2008). PEST is a model independent non-linear parameter estimator. Since its inception in the mid 1900s, PEST has become the industry standard in the calibration of all kinds of environmental modelling problems (Doherty and Johnston, 2003). PEST automatically minimizes the objective function (sum of squared errors) using the Gauss-Marquardt-Levenberg optimisation algorithm. BEOPEST (Schreuder, 2009), A PEST utility program was used to in this study to run PEST in parallel computers. BEOPEST uses a master and slaves that communicate via TCP/IP. Using BEOPEST, 10 personal computers were run in parallel during the calibration. The calibrated degradation rate constants for PCE, TCE, DCE, and VC respectively are: 0.0021, 0.0033, 0.0044, and 0.0032 day^{-1}. The calibrated first-order degradation rates are in the range of reaction rate constants given in Wiedemeier et al. (1996). Based on lab and field-scale results from previous studies Wiedemeier et al. (1996) reported that the representative first-order biodegradation rates for chlorinated solvents range from 0.00068 to 0.54 day^{-1} for PCE, 0.0001 to 0.021 day^{-1} for TCE, 0.00016 to 0.026 day^{-1} for DCE, and 0.0003 to 0.012 day^{-1} for VC. The mean values of the observed and simulated concentration values obtained from the concentration time series used in calibration are presented in Table 3.3. The mean absolute error is less for VC than DCE in both layers. Given the limited amount of data available for calibration and the assumption of constant concentration at the source the model can provide reasonable approximation that can be used for scenario analysis.

Table 3.3: calibration statistics (observed versus simulated) performance measures at observation well SB1.

Description	Mean of observed (μg/L)	Mean of simulated (μg/L)	Mean absolute error (μg/L)
DCE layer1	877	705	172
DCE layer2	308	549	240
VC layer1	1949	1921	28
VC layer2	1296	1518	223

3.6   Results and Discussions

3.6.1   Baseline Scenario

Figure 3.7 and Figure 3.8 show the concentration distribution in layer 1 and layer 2, 21137 days (November 2007) after the simulation start. The PCE concentration distribution in the second layer is wider and longer than the plume in the first layer. The TCE concentration distribution in the second layer shows higher concentration, contours are moving away from the source and the plume is getting wider. The DCE and VC concentration contours shown in Figure 3.8 show similar trend. The concentration contours are getting wider and moving faster towards the receptor. This is mainly due to the high hydraulic conductivity in the second layer.

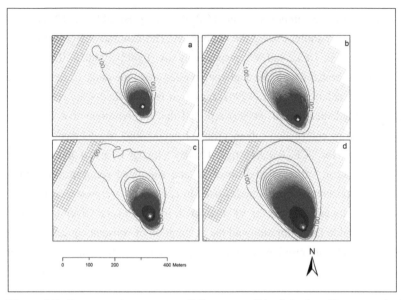

Figure 3.7: Concentration contours (Micrograms/Litre) for baseline scenario, PCE layer 1 (a) PCE layer 2 (b), TCE layer 1 (c), and TCE layer 2 (d) (contour interval 1000 µg/L).

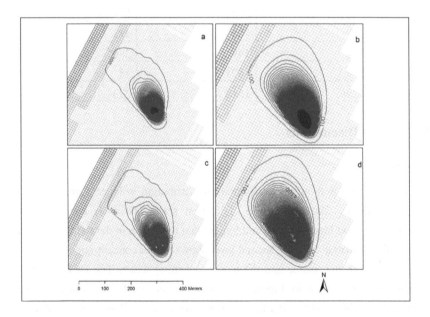

Figure 3.8: Concentration contours (Micrograms/Litre) for baseline scenario, DCE layer 1 (a) DCE layer 2 (b), VC layer 1 (c), and VC layer 2 (d) (contour interval 1000 µg/L).

3.6.2   Source Removal and natural attenuation Scenario

DNAPL zones in the aquifers serves as a long term continuous source after the source loading was discontinued. In this study, we investigated the role of source removal from the saturated zone in reducing contaminant concentration down gradient of the source. Using baseline scenario, simulation was completed till end of 2013. The 2013 concentrations were then used as initial concentrations for the next predictive simulation with source removal scenario, till 2035 (23 years of predictive simulation after source removal). The source area considered during the baseline scenario was 20 m x 20 m, but for this scenario we considered source removal of 100 m x 100 m of source zone area. The objective is to evaluate the natural attenuation capacity of the aquifer after the source zone contaminant concentrations are successfully reduced by removing the source. This was done by comparing the simulation results with the remediation standards. Groundwater remediation standards for Flanders were obtained from Bronders et al. (2007). The remediation standard respectively for PCE, TCE, DCE and VC are: 40, 70, 50 and 5 µg/L.

Figure 3.9 shows simulation of PCE and TCE concentration along the profile line from the source for both natural attenuation scenario with source removal, and natural attenuation only scenario. As it can be seen in Figure 3.9 the concentration of PCE and TCE down gradient is decreasing for both scenarios below the remediation standard. However, it seems that removing the source only help to reduce the concentration of these chemicals close to the source. Similar observation observations can be made from Figure 3.10 for DCE and VC concentration along profile line. The difference between the two scenarios can be seen clearly in this case. Unlike the parent chemical (PCE and TCE), source removal has significant impact on the concentration of the daughter products (DCE and VC). For instance, VC concentration using source removal and natural attenuation is 21 % lower than natural attenuation only scenario at the end of the simulation time moreover, the percentage reduction is much higher the first 250 m. However, for both scenarios considered the concentration of VC is much higher than the remediation standard.

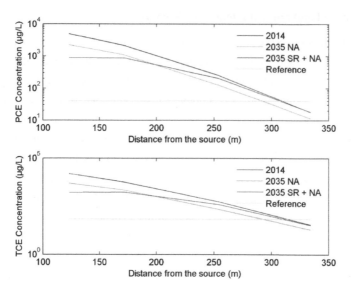

Figure 3.9: Simulation concentration profile of PCE and TCE along Transect A-B corresponding to the year 2014 and 2035 with natural attenuation only (NA) and source removal and natural attenuation (SR+NA), y-axis in semi log scale and Reference represent groundwater remediation standard.

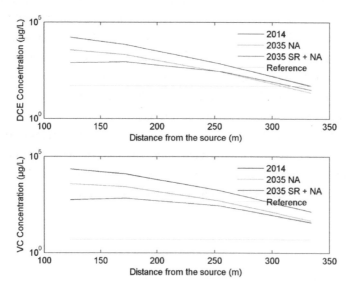

Figure 3.10: Simulation concentration profile of DCE and VC along Transect A-B corresponding to the year 2014 and 2035 with natural attenuation only (NA) and source removal and natural attenuation (SR+NA), y-axis in semi log scale and   Reference represent groundwater remediation standard.

3.6.3   Sensitivity Analysis

A sensitivity analysis was performed to investigate the influence of the degradation rate constants. Parameter variation is usually expressed as a percentage change from the calibrated model value. The magnitude of the change in concentrations from the calibrated values is a measure of the sensitivity of that particular parameter. In this study, the reaction rates were perturbed 50% above and below the calibrated values (i.e. the baseline), and the resulting concentration profiles are shown in Figures 3.11 and Figure 3.12. Sensitivity analysis was carried out for predictive simulation which corresponds to the year 2035 starting from 2014. Predictive simulation was carried out to estimate contaminant concentration as time progresses. Such concentration information is the primary metric for determining the success of the remediation scenario like natural attenuation. Results of the sensitivity analysis showed that DCE and VC are more sensitive to degradation rates than the parent chemicals. However, the sensitivity analysis results revealed that increasing the degradation rate by 50% could reduce the VC concentration close to the receptor location, but not to the remediation standard. The sensitivity of the daughter products to degradation rate were observed to increase with distance from the source.

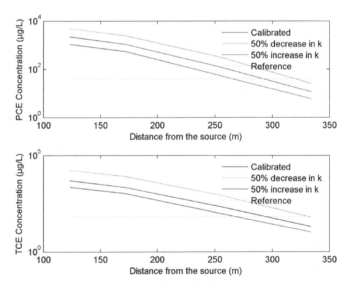

Figure 3.11: Sensitivity of PCE and TCE profiles along Transect A-B corresponding to the year 2035 for scenario natural attenuation, y-axis in semi log scale and Reference represent groundwater remediation standard.

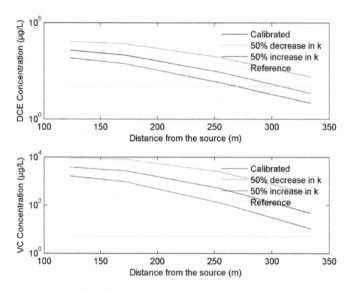

Figure 3.12: Sensitivity of DCE and VC profiles along Transect A-B corresponding to the year 2035 for scenario natural attenuation, y-axis in semi log scale and Reference represent groundwater remediation standard.

3.7 Conclusions

A multi-species reactive transport model was applied at chlorinated solvent study site in Vilvoorde-Machelen. The flow and transport models were calibrated to reflect the field conditions observed at the site. The calibrated degradation constants are within the range of values reported in literatures. However, given the limited amount of data used for calibration, the calibrated parameters are by no means unique solutions.

Sensitivity analysis was carried out to identify the sensitivity of first order degradation rate. Results of the sensitivity analysis indicate that the parent chemicals (PCE and TCE) are less sensitive to degradation rates while the degradation products (DCE and VC) are more sensitive. The concentration profile lines are wider as the plume moves to the receptor location. Moreover, it was observed that the concentration plumes are wider and moving faster towards the receptor in the high hydraulic conductivity zones. Predictive simulations 13 years after the calibration period showed that only VC cannot degrade to the level recommended by groundwater remediation standard. Therefore for this chemical natural attenuation may not be good remediation option. However, the history modelling showed that others (PCE, TCE and DCE) are already below remediation standard.

4 Local Scale Groundwater-Surface water Interactions in Modified Urban Rivers

4.1 Introduction

Numerical groundwater models such as MODFLOW have been widely used for exploring groundwater-surface water interactions either by representing surface water features within the model (e.g. river, lake packages) or by their coupling with surface water models. Although these numerical models are valuable tools for understanding the dynamics of interaction, they are not easily applied where hydraulic head data may be sparse or unevenly distributed. It is very difficult economically to obtain data on hydraulic heads with enough horizontal and vertical density for a particular groundwater region. A recent study by Brunner et al. (2010) examined the applicability of the MODFLOW model to simulate groundwater-surface water interactions for connected and disconnected losing streams. These authors have identified four major limitations of MODFLOW: (1) the inability to simulate the negative pressure beneath disconnected streams, resulting in underestimation of the infiltration flux; (2) a river may be either fully connected or disconnected, however, in reality transitional stages between the two flow regimes may exist; (3) a mismatch between river width and the grid cell results in an error because the exchange rates between surface water and groundwater are distributed over the area of the grid cell; (4) an error occurs in simulating the height of the groundwater mound by using a coarse vertical discretisation so as to avoid drying out of cells. On the other hand, the use of heat as natural tracer in conjunction with water level measurements for quantifying groundwater-surface water interactions has proven to be an effective method (Anderson, 2005; Stonestrom and Constantz, 2003). However, the type of interactions between stream temperature, stream flow, and groundwater depend on whether the stream is losing or gaining in a given reach (Constantz, 1998).

Numerical models of coupled groundwater flow and heat transport models are usually calibrated with observed data and vertical flux is estimated by inverse modelling with estimation of thermal and hydraulic parameters. A number of studies have explored the effects of spatial and temporal variability of groundwater-surface water flux exchanges (e.g. Keery et al. (2007), Lowry et al. (2007), Schmidt et al.(2006)). However, little attention has been given to the impact of temporal resolution of the input data on model calibration. The usual practice in the model calibration is to set the length of the time step interval to the sampling interval. In most cases, the temporal resolution is fixed by the data collection procedure. For instance, if a modeller has available daily data, a daily model can then be calibrated by matching model simulations to the actual daily time series. The usual approach is to assume that the model parameters are constant within the sampling interval or model calibration time step. A fundamental issue associated with this modelling approach is that, in reality, calibrating the model with long-term input data may hide the information on processes that have short-term responses. Moreover, several researchers in the field of hydrology (e.g Littlewood and Croke (2008), Littlewood et al. (2011), Wang et al. (2009), Wang et al. (2011), Wetterhall et al. (2011)) demonstrated the sensitivity of hydrological models to the temporal resolution of the input data. To assess the significance of this issue, this study aims to examine the effect of temporal resolution of input data on temporal variation of groundwater-surface water flux exchange. To this end, the Zenne River (Belgium), which is an extensively modified urban river, was used as a case study. A vertical two dimensional groundwater flow and heat transport model was calibrated at hourly and daily time scales using water level and temperature data observed in the river and the adjacent aquifer, as well as temperatures measured at multiple depths in the streambed for simulating the flux exchange between the adjacent aquifer and the river. The daily data used in this study are data averaged from the measured hourly observations.

4.2 Study Area

The study site is located at Vilvoorde-Machelen, described in chapter 3. The focus of this study is cross section A-A (Figure 4.1). Due to more than a century of industrialisation the site has been polluted to the extent that individual plumes are not definable any more (Bronders et al., 2007). Field investigations on the study area indicate the presence of chlorinated aliphatic hydrocarbons (CAHs) in the regional groundwater. A groundwater plume contaminated with PCE and TCE (originating from several sources) flows towards the Zenne River. Hamonts et al. (2009) reported that those pollutants are reductively dechlorinated in the plume and only the products of this process, i.e., vinyl chloride (VC), and cis-dichloroethene discharge into the Zenne River at its right riverbank while unpolluted groundwater discharges at the opposite riverbank. In addition to subsurface pollution from multiple sources, the Zenne River received domestic sewage at various locations until January 2007, which created highly eutrophic conditions in the surface water and the riverbed (Hamonts et al., 2009). Along the bank of Zenne River there is a 9 m deep steel sheet pile wall driven through the aquifer. At the study site the Zenne River is relatively straight, 12-15 m wide, 0.5-2 m deep and has a stream flow of 5-10 m^3/s in dry weather conditions. Bed materials range from fine to coarse grained sand and gravel to silt and slick. The lithological graphic section from the adjacent aquifer at well SB2 is shown in Figure 4.2.

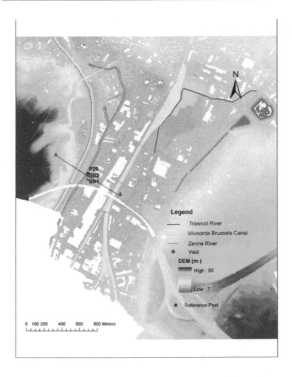

Figure 4.1: Digital elevation model image of the study area.

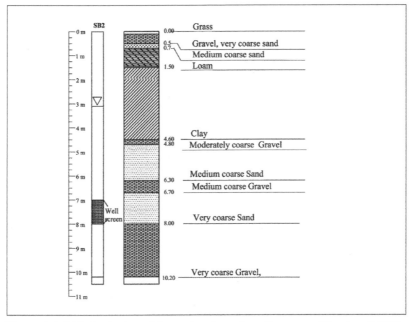

Figure 4.2: Lithological graphic section from the adjacent aquifer at well SB2.

4.3   Water Level and Temperature Data

The groundwater level and temperature data used in this study were obtained from the Flemish Institute for Technological Research (VITO). VITO has performed various monitoring campaigns at the test site, particularly for studying the reductive dechlorination of chlorinated solvents (Hamonts et al., 2009). Temperature and water level data were collected every 30 min in groundwater well SB2 (screened at 7-8 m below the ground surface) using a MiniDiver installed in the well. This well is located 32.5 m upstream of reference post 26, at 3.5 m from the right riverbank. Temperature and water level data for the Zenne River surface water were collected with similar sensors, installed at the top of the riverbed within a filtered protection pipe, near the right riverbank in the test area, 24.5 m upstream of reference post 26. Pore water temperatures below the riverbed were collected using temperature sensors that were installed at 20, 60 and 120 cm depth in the riverbed 24.5 m upstream of reference post 26. Every 30 min, the pore water temperatures were stored by a data logger that was connected to the sensors and installed inside a field housing at the riverbank. Figure 4.3 shows the observed water level for the Zenne surface water and groundwater level at well SB2. Except during high flow periods, the groundwater level in well SB2 is higher than the surface water level of the Zenne River.

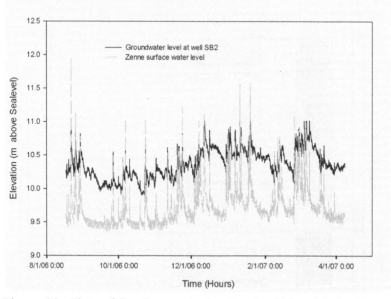

Figure 4.3: Plots of hourly measured water level data of the Zenne River and groundwater at well SB2 from August 2006 to April 2007.

4.4   Methods

Understanding of contaminant flux at the river-aquifer interface requires detailed information on the magnitude and temporal variability of water flux exchanges across this interface. In this study, a flow and heat transport model was used to quantify the temporal water flux exchange at the river-aquifer interface. The produced model broadly captures the timing and magnitude of the flux exchange encompassing a storm event. The following subsections describe the numerical model, model set-up, model input, model calibration and validation.

4.4.1   Numerical Model

VS2DH was used to simulate flow and energy transport across the streambed. For this study, relations between pressure head, moisture content, and relative hydraulic conductivity are represented by functions developed by van Genuchten (1980).

4.4.2   Model Set-up

Spatial Discretisation

A two-dimensional model is set up to represent a vertical slice of the streambed. The cross-section of the model setup used in the numerical simulation is shown in Figure 4.4. Because of the lack of water level and temperature data at the left bank of the river, only half of the river cross-section was considered for simulation. The 2-D simulation grid was constructed with 101 rows and 55 columns. The grid sizes of the rows were varied from 0.06 m at the riverbed to 0.2 m at the bottom while a uniform column width of 0.2 m was used.

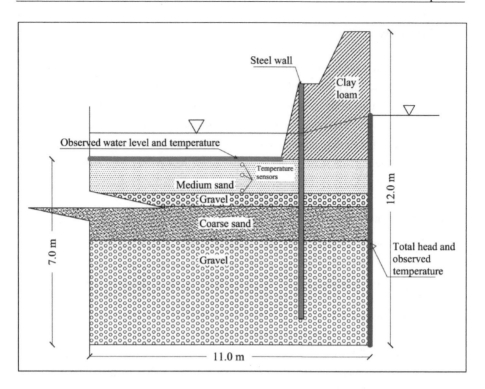

Figure 4.4: Boundary conditions and temperature sensors installed at 20, 60, and 120 cm depth in the riverbed for the VS2DH model for the Zenne River cross-section.

Initial Conditions

For the initial hydraulic condition, initial pressure distribution in the unsaturated part of the region of interest was specified as 'initial equilibrium profile', which is defined as the pressure head being in equilibrium with the negative elevation head above a free water surface or water table. These initial conditions usually represent some type of steady state or equilibrium. In the saturated part, measured groundwater heads were specified as initial conditions. For the heat transport, the initial temperature field needs to be specified. The initial temperature for the model was determined by drawing temperature contours from the observed temperature data. Therefore, three contours were drawn in the model domain: at the top boundary in the riverbed from the observed streambed temperature, at the deepest sediment layer from the observed sediment pore water temperature and at the bottom boundary using the temperature observed in the deepest groundwater observation well SB2. These observed temperature values were interpolated by the model in the vertical direction to create an initial geothermal profile.

Boundary Conditions

In addition to initial conditions, boundary conditions need to be specified along the boundaries of the model domain. In this case, the middle of the river section was chosen as a line of symmetry and used as no flow boundary condition. This assumption is consistent with other studies (Stonestrom and Constantz, 2003; Su et al., 2004). The right hand side was simulated with a variable total head boundary condition. Surface boundaries representing the stream-subsurface interface were assigned as a time-varying observed temperature and specified pressure head. The measured temperature and water level data from August 2006 till April 2007 were used as boundary condition at the stream-groundwater interface. The bottom boundary is an interface between the saturated, pervious soil mass and adjacent soil with a very low permeability which is approximated as an impervious boundary. It is assumed that no flow takes place across this interface, thus flow in the pervious soil next to the impervious boundary is parallel to that boundary. Temperature at the lower boundary was linearly extrapolated from data available at observation well SB2.

Temporal Discretisation

The model domain was simulated using increasingly finer discretisation both in space and time. The discretisation was centred in space and time. An initial time step of 0.001 h and adaptive time step constrained at a minimum and maximum time step of 0.001 h and 1 h were used for the hourly model. Similarly, the daily model was run with an initial time step of 0.001 d constrained at minimum and maximum time step of 0.001 d and 1 d. An adaptive time step factor of 1.2 was used for both models; hence the time step size was increased for each successive time step by a factor of 1.2 until a maximum time step was reached. The total hourly and daily transient flow and transport model simulations consist of 5614 one-hour and 234 one-day recharge periods respectively.

4.4.3   Model Input

The VS2DH model needs as inputs estimations of hydraulic and thermal properties such as porosity, heat capacity of dry sediment, thermal conductivity of the saturated sediment, horizontal hydraulic conductivity, anisotropy ratio, and horizontal and longitudinal dispersion. Therefore, different textural classes were used to define the flow and transport parameters. The porosity of the streambed sediment and bulk densities were determined by Hamonts (2009). Thermal conductivity and heat capacity of the sediment were estimated from published values (Lapham, 1989). Lapham (1989) discussed various ways to determine the physical and thermal properties of the sediment. He reported that the inexpensive way of determining the physical and thermal properties of saturated fine and coarse grained sediment is to relate laboratory determined wet and dry-bulk densities to physical and thermal properties of the sediment. He has published ranges of values for heat capacity, thermal conductivity and diffusivity of saturated fine-grained and coarse grained sediment, and various graphs that show the relation between heat capacity, thermal conductivity, and diffusivity of fine and coarse grained sediment to sediment dry bulk density. Thermal properties used in this study are in agreement with other studies carried out in the Aa River in Belgium (Anibas et al., 2009).

The hydraulic conductivity was determined through calibration. Hydraulic conductivity shows high variability as compared to the thermal conductivity of the same material (Stonestrom and Constantz, 2003). For instance, various studies showed that stream sediments composed of sand and gravel can have a hydraulic conductivity that is six orders of magnitude higher than clay. However, the thermal conductivity ranges between 2.2 W/m °C (coarse grained sand) and 1.4 W/m °C (clay). Therefore, the common approach is to assume the thermal parameters based on literature values for a particular textural class as they can be specified within a narrow range. Temperature distribution in the streambed is more sensitive to the hydraulic conductivity (a parameter that varies over several orders of magnitude), which controls the flux.

To represent the effect of the steel sheet pile wall on the heat transfer and flow processes, a textural class of porous medium with a very low hydraulic conductivity was used (Richard W. Healy, personal communication, 2011). Although the steel sheet piles are impermeable, the interlock joints between adjacent sheet piles are typically permeable, which allows leakage through these joints (Sellmeijer et al., 1995). The flow through the porous medium used to represent the interlock joints of the steel sheet pile wall is governed by Darcy's Law. For a given steel sheet pile wall the properties of porous material with the same seepage properties can be calculated. Assuming a porous material wall of thickness d (m) and a horizontal distance between the joints in the pile b (m), the equivalent hydraulic conductivity, k_e (m/s) along the horizontal direction can be computed by ($k_e = \rho*d/b$). Where, ρ is the inverse joint resistance (m/s), which depends on the type of sealing/filler material, the soil, the driving method and the interlock. On the other hand, the thermal conductivity of the steel wall is about 30 times larger than that of the soil material. The heat flow between the steel wall and the soil is thus relatively small compared to the heat flow along the steel sheet pile itself. Therefore, a thermal conductivity value for the steel wall was calculated from literature by equating the conductive heat transfer using Fourier's Law. A similar calculation for the heat capacity of the porous material was made using the relation between heat capacity and the total amount of heat required for heating a material from one temperature level to another.

4.4.4   Model Calibration

Model calibration was completed for both hourly and daily time scales. In order to calibrate the model with respect to the observed sediment temperature at three vertical depths, the VS2DH model was linked to PEST. PEST was used in this study to minimize the sum square error of observed and simulated temperatures (Equation 4.7).

$$Min\ \Theta = \sum_i (\dot{T}_i - T_i)^2 \tag{4.7}$$

Where Θ is the objective function to be minimized $\dot{T}_i$ is measured sediment temperature ($^{\circ}$C), and T_i is simulated sediment temperature ($^{\circ}$C).

Temperatures observed at three depths for a period of 158 days (August, 17, 2006 till January, 22 2007) was used for model calibration. All the three observation points were assumed to have equal weight in the automatic calibration process. Simulated temperature values were matched with observed sediment temperature values by optimizing the hydraulic and thermal parameters. Figure 4.5 shows the observed and simulated sediment temperature at three different depths during the hourly model calibration. Similarly, Figure 4.6 shows the observed and simulated sediment temperature at three depths during the daily model calibration. The obtained results show a very good match between the observed and simulated sediment temperature values. Generally the attempt was to reproduce the phase (timing) and amplitude of the subsurface temperature variations. The patterns of the simulated and observed temperatures are also very similar (Figure 4.5 and 4.6). Tables 4.1 and 4.2 summarise the hydraulic and thermal parameters used in the final calibrated models. In order to evaluate the magnitude of difference between observed and simulated temperature values at 20, 60, and 120 cm depth below the riverbed, several performance measures were used as presented in Table 4.3. Overall, the performance measures show that both models are reasonably well calibrated for the intended purpose.

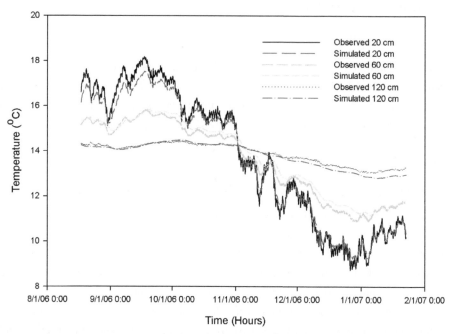

Figure 4.5: Observed streambed temperature compared with simulated streambed temperature at 20, 60, and 120 cm depth below the riverbed hourly model calibration.

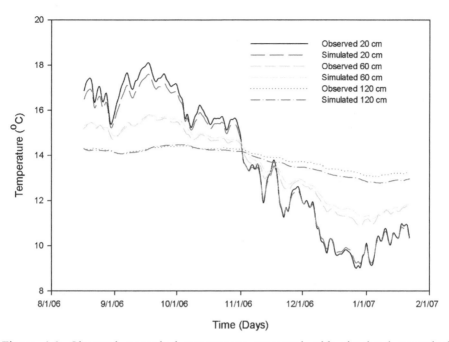

Figure 4.6: Observed streambed temperature compared with simulated streambed temperature at 20, 60, and 120 cm depth below the riverbed, daily model calibration.

Table 4.1: Summary of optimized hourly model parameters used in the hourly VS2DH model.

Parameter	Medium Sand	Coarse Sand	Gravel	Clay Loam	Steel wall
Hydraulic conductivity (m/d)	0.08	0.42	1.33	0.11	0.01
Porosity	0.38	0.390	0.42	0.43	0.2
Heat capacity of dry solid (J/m^{3o}C)	2.70×10^6	2.00×10^6	2.00×10^6	2.70×10^6	2.00×10^5
Thermal conductivity (J/d/m ^{o}C)	129600	172800	172800	155520	371520
Longitudinal dispersivity (m)	0.10	0.1	0.50	0.10	0.1
Transverse dispersivity (m)	0.10	0.1	0.10	0.10	0.1
Heat capacity of water (J/m^{3o}C)	4.20×10^6				

Table 4.2: Summary of optimized daily model parameters used in the daily VS2DH model.

Parameter	Medium Sand	Coarse Sand	Gravel	Clay Loam	Steel wall
Hydraulic conductivity (m/d)	0.116	0.416	1.33	0.107	0.005
Porosity	0.375	0.390	0.42	0.43	0.2
Heat capacity of dry solid (J/m^{3o}C)	2.70 x10^6	2.00x10^6	2.00x10^6	2.70 x10^6	2.00 x10^5
Thermal conductivity (J/d/m ^{o}C)	129600	172800	172800	155520	371520
Longitudinal dispersivity (m)	0.10	0.1	0.50	0.10	0.5
Transverse dispersivity (m)	0.10	0.1	0.10	0.10	0.1
Heat capacity of water (J/m^{3o}C)	4.20x10^6				

Table 4.3: Comparison of observed and simulated temperature time series using different performance measures during the calibration period August, 17, 2006 till January, 22, 2007

	Hourly model calibration			Daily model calibration		
Depth below riverbed (cm)	20	60	120	20	60	120
RMSE[1] (^{o}C)	0.32	0.23	0.14	0.25	0.18	0.18
Cor.[2]	0.99	0.99	0.99	0.99	0.99	0.99
MAE[3] (^{o}C)	0.26	0.19	0.11	0.21	0.15	0.15
STD[4] of observed temperature (^{o}C)	2.94	1.65	0.45	2.95	1.65	0.45
STD[4] of Simulated temperature (^{o}C)	2.72	1.61	0.55	2.76	1.58	0.57
Mean of observed temperature (^{o}C)	13.75	13.60	13.90	13.76	13.60	13.90
Mean of simulated temperature (^{o}C)	13.62	13.78	13.81	13.60	13.74	13.76

[1] RMSE (Root mean square error) between the measured and simulated temperature (^{o}C)

[2] Cor.(correlation) between the measured and simulated temperature

[3] MAE (Mean absolute error) a measure of the average difference between observed and simulated temperature (^{o}C)

[4] STD is standard deviation of observed and simulated temperature (^{o}C)

4.4.5   Model Validation

In order to check the validity of the calibrated hydraulic and thermal parameters, the hourly and daily models were validated with an independent data set from January, 23, 2007 till April, 8, 2007. For measuring the performance of the model, a standard set of criteria such as graphical plots (Figure 4.7 for the hourly model and Figure 4.8 for the daily model) as well as numerical measures (Table 4.4) were used. As shown in Table 4, the performance of the model along the three depths is reasonably good. In general, the validation results for both the hourly and daily time scales show that the model performance was reasonably good in simulating temperature time series for periods outside of the calibration period.

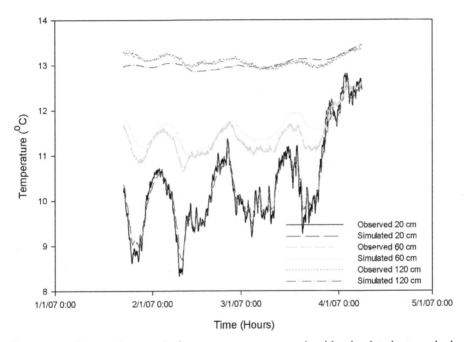

Figure 4.7: Observed streambed temperature compared with simulated streambed temperature at 20, 60, and 120 cm depth below the riverbed, hourly model validation.

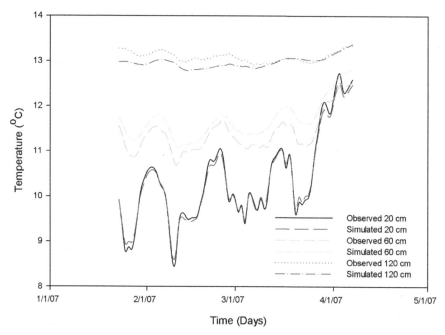

Figure 4.8: Observed streambed temperature compared with simulated streambed temperature at 20, 60, and 120 cm depth below the riverbed, daily model validation.

Table 4.4: Comparison of observed and simulated temperature time series using different performance measures during the validation period January, 23, 2007 till April, 8, 2007.

	Hourly model validation			Daily model validation		
Depth below riverbed (cm)	20	60	120	20	60	120
RMSE ($^{\circ}$C)	0.17	0.27	0.13	0.08	0.30	0.16
Cor.	0.99	0.96	0.50	0.99	0.98	0.61
MAE ($^{\circ}$C)	0.14	0.24	0.11	0.07	0.28	0.13
STD of observed temperature ($^{\circ}$C)	1.02	0.44	0.11	1.00	0.43	0.11
STD of Simulated temperature ($^{\circ}$C)	0.91	0.42	0.13	0.94	0.42	0.13
Mean of observed temperature ($^{\circ}$C)	10.37	11.41	13.10	10.36	11.40	13.09
Mean of simulated temperature ($^{\circ}$C)	10.38	11.65	13.04	10.34	11.68	12.97

Abbreviations are the same as in Table 4.3

4.5 Results and Discussion

4.5.1 Effect of Temporal Resolution on Model Parameters and Performance

In order to investigate the effect of temporal resolution on model parameters and performance, two numerical model calibrations were conducted. In the first calibration, daily water level and temperature data averaged from the hourly values were used as input. In the second calibration, hourly water level and temperature data were used as input. Different model parameters were then obtained during the two calibration experiments as shown in Table 4.1 and 4.2. In the optimisation, initial values for the time-dependent parameters of the hourly model were determined by dividing the calibrated daily parameters by 24. Research in the field of hydrology (Littlewood and Croke, 2008; Littlewood et al., 2011) demonstrated that both time dependent and non-time dependent parameters could change with the model time step. In this study, we also found that model parameters that have high control on flux exchange (hydraulic conductivity) could change during the two calibration time scales. We have used PEST sensitivity analysis to identify the most sensitive parameters for the hourly time scale. We have done this by dividing the daily calibrated parameters by 24. The hydraulic conductivity particularly below the riverbed (medium sand) was found to be the most sensitive parameter to the hourly observations.

Streambed hydraulic conductivity is affected by the direction and magnitude of the vertical gradient within the streambed. As reported by Hannula and Poeter (1995) and more recently by Rosenberry and Pitlick (2009), the downward gradient increases the effective stress which reduces the size of the pore openings in the streambed material, thereby lowering the hydraulic conductivity. On the other hand, an upward gradient may reduce the effective stress, which may increase the hydraulic conductivity. Therefore, even when the sediment conditions are constant, the hydraulic conductivity at a point on the streambed may be higher when an upward gradient prevails than when a downward gradient occurs. During the flood events the water level in the river rises quickly and exceeds the water table in the adjacent aquifer. Such events may also induce riverbed scouring during the rising limb of the hydrograph, which temporarily leads to increased vertical hydraulic conductivity. However, this hydraulic conductivity can subsequently decrease after storm events due to deposition of suspended load that can clog the pore spaces (Simpson and Meixner, 2012). The results of the optimised hydraulic conductivity values for the layer immediately below the riverbed seem to indicate that the hourly model is capable to capture some of these conditions (with k=0.08 m/d and lower than k=0.116 m/d, obtained from the daily model). The hourly model captures better the short events with downward gradients, which is not the case for the daily model. During these events increased effective stress may lead to lower hydraulic conductivity values, while in the periods between them lowering of hydraulic conductivity may be caused by deposition of suspended sediments. The scouring effect that may lead to short term increase in hydraulic conductivity during these events seems to be small (the duration of the rising limb in these events is always very small) and dominated by the above described processes that lead to lowering of hydraulic conductivity. For the other layers included in the model such differences between the hourly and the daily model are insignificant, which indicates that the processes described above indeed are mostly related to the layer just below the riverbed.

Error statistics, including mean absolute error (MAE) and root mean square error (RMSE) for the daily and hourly models at 20, 60, and 120 cm depth below the riverbed are shown in Table 4.3. Due to the smoothing effect of the daily averaging, the model was more successful at matching the daily temperature compared to the hourly temperature. Another interesting issue is the relative predictive power that can be expected at the two different time scales. The models' predictive power was assessed by comparing model performance in the validation period (Table 4.4). Except for the temperature at 20 cm depth, the predictive performance was better for the hourly model.

4.5.2   Effect of Temporal Resolution on Flux Exchange

Flux values were calculated as flow across the model boundaries in m^3/h and m^3/d respectively. Since we have specified the streambed as specified head boundary, the flux out of this domain across this boundary was extracted from the mass balance summary, as it represents the flux across the streambed. The flux across the boundary was extracted for the total length of the boundary, which varies along the boundaries. The flux per unit area was calculated by dividing the total flux across the boundary by the boundary length. In this way the flux per unit area was calculated for both hourly and daily models. Result of the analysis showed that the streambed flux including the calibration and validation period, ranged from -37 mm/d to +25 mm/d using the hourly model (positive indicate flow is from surface water to groundwater) (Figure 4.9 a and 4.9 b) and from -10 mm/d to -37 mm/d using the daily model (Figure 4.10). The continuous flux simulation demonstrates the dynamic nature of groundwater-surface water interactions. Overall, the continuous flux simulation using the daily time scale model shows less dynamics than the hourly model. An inherent assumption was that the use of an hourly time step model would allow to quantify surface water infiltration events that may occur due to rapid rises in river water level and would thereby prove the occurrence of dilution of the discharging groundwater contaminants by surface water-mixing in the river sediments and thus result in a better evaluation of the CAH attenuation efficiency of the river sediments than the daily time scale model. The reason for this assumption was the fact that during high flow events the hourly observed data revealed periods when the river water level was higher than the groundwater table in the aquifer (see in Figure 4.9a and 4.9b for the graph of water level difference between well SB2 and the Zenne River). Such periods were not revealed with the daily averaged inputs (see corresponding graph in Figure 4.10). The results in terms of obtained flux show that the exchange of water is always from the aquifer to the river in the daily model whereas in the hourly model, indeed, transient reversal in flow from the river to the aquifer was observed. The rapid rise in river stage compared to the hydraulic head in the underlying aquifer causes a reversal in head gradient, and river water is pushed into the underlying aquifer. For the period considered in this study there were 12 short period reversals in flow (induced by high stream flow events) that coincided with rapid spikes in the Zenne River water level shown in Figure 3. A summary of the flux from the river to the aquifer and duration of flow reversal from the river to the aquifer is provided in Table 4.5. As shown in Table 4.5, the flux from the river to the aquifer ranged from 0.1 mm/d to 25 mm/d and the duration of flow reversal ranged from 1 h to 6 h. Maximum infiltration to the underlying aquifer occurs in the month of August 2006 for the duration of 6 h.

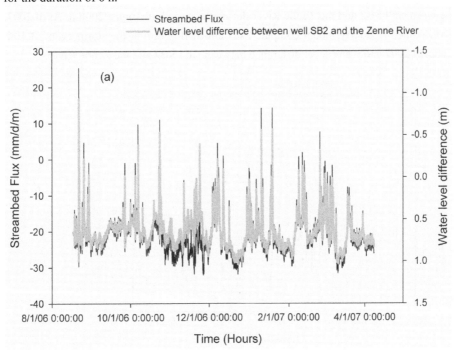

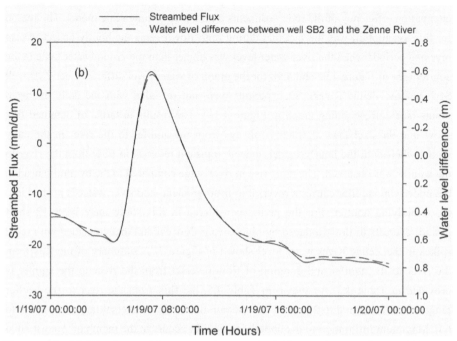

Figure 4.9: (a) Hourly model computed streambed flux and water level difference between well SB2 and the Zenne River during the period of August 2006 to April 2007. (b) Zoomed-in hourly model computed streambed flux during the storm event of 19th January 2007 and water level difference between well SB2 and the Zenne River.

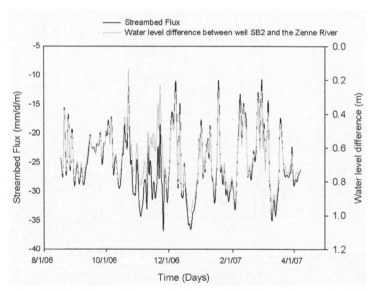

Figure 4.10: Daily model computed streambed flux and water level difference between well SB2 and the Zenne River during the period of August 2006 to April 2007 (negative numbers indicate flow from the aquifer to the river).

Table 4.5: Summary of fluxes from the river to the aquifer and duration of flow reversals

Duration	Flux (mm/d)			Duration of flow reversal (h)
	Minimum	Maximum	Average	
8/21/2006 18:00- 23:00	7.5	25	16.59	6
8/25/2006 15:00			4.5	1
10/6/2006 22:00 -23:00	5.5	7.9	6.69	2
10/23/2006 16:00-23:00	0.1	9.3	4.20	3
11/23/2006 16:00			0.38	1
12/7/2006 15:00-16:00	1.2	4.5	2.84	2
12/12/2006 3:00-4:00	0.3	0.7	0.47	2
1/10/2007 18:00-20:00	4.9	13	8.82	3
1/19/2007 6:00-9:00	0.2	13.8	6.77	4
2/8/2007 17:00			0.7	1
2/11/2007 11:00-12:00	0.2	0.5	0.36	1
2/25/2007 6:00 - 8:00	1.9	7.4	4.09	3

The results obtained in this study are in agreement with the MODFLOW results obtained in chapter 3 and results from previous studies for the Zenne River. In Dujardin et al. (2011) the estimated groundwater flux to the river using the steady state groundwater model and heat transport model was 34 mm/d and 35 mm/d respectively. The steady state groundwater model was calibrated using average groundwater level data from 1999 to 2006. Therefore this model could only represent the average flux from the groundwater to the stream but it could not capture the temporal variation in flux estimates. The heat transport model used was also steady state. It only represented the flux estimates of one week for the month of April 2006 as well as October-November 2006. Anibas et al. (2009) found that the flux obtained from the steady state solution is in agreement with estimates from the transient case only for certain time periods, whereas for other periods they could differ significantly. Moreover, the study by Dujardin et al. (2011) did not take into account the effect of the steel wall on the flux estimation.

4.5.3 Implications of Flux Exchange Modelling for Contaminant Transport

Similarly to many urban rivers, the Zenne River responds rapidly to rainfall events as a result of significant runoff generation and a rapid routing of urban flows through urban drainage networks. Knowledge of the flow systems at the interface between groundwater and surface water is a critical component towards establishing site-specific remedial alternatives for contaminant transport. Water flow dynamics influence contaminant transport due to changes in the volume and the direction of the water flux. Therefore, information on temporal variation of groundwater-surface water flux exchange is essential to support any quantified understanding of contaminant attenuation capacity in the hyporheic zones. For instance, calculated temporal variation in the groundwater- surface water flux exchange provides detailed temporal information on the residence time of the groundwater in the riverbed, which can be used to evaluate the biotransformation efficiency of contaminants in the hyporheic zone over time.

Based on the flux calculated in this study, a groundwater residence time between 120 and 20 cm depth in the Zenne riverbed sediments of at least 31 days would allow the observed extent of biotransformation of VC in the Zenne River sediments (Hamonts et al., 2012) based on the half-life for VC of 6 to 26 days obtained from microcosm tests (Hamonts, 2009). On the other hand, surface water infiltration into the riverbed during high precipitation events contributes to dilution of CAHs in the Zenne riverbed. A previous study showed that dilution by surface water-mixing often contributes more to the total CAH reduction than the observed biotransformation in the Zenne River sediments (Hamonts et al., 2012). Therefore, quantifying both biotransformation and dilution and monitoring the extent of these process over time was shown to be essential for a correct evaluation of the efficiency of CAH attenuation in river sediments receiving a CAH-polluted groundwater plume (Hamonts et al., 2012).

4.6 Conclusions

This study investigated the effect of temporal resolution of water level and temperature data on the simulation of water flux exchange between the Zenne River and its adjacent aquifer using VS2DH, a groundwater flow and heat transport model calibrated with multiple depth temperature measurements at hourly and daily time scales. Temperatures observed at multiple depths in the sediment at the river-aquifer interface were adequately reproduced by both models. Results of the study showed that the hourly model better simulated the dynamics of the river-aquifer exchange particularly during storm events. The river stage increase during these storm events caused a reversal of the Zenne River from gaining to losing conditions in the hourly model, whereas the daily model suggested the existence of flow from the groundwater to the river all the time. Application of the method at the test site demonstrates the effect of temporal resolution of water level and temperature measurements on temporal variability of groundwater– surface water flux exchanges. The model results can therefore be used for quantifying the contaminant flux flowing from the groundwater to the river as well as to support any remedial alternatives. As with any modelling effort, there are uncertainties associated with the analysis: (i) the assumption that the flow is symmetric in the middle of the river cross-section may not be verified from two observation wells (SB1 and SB2) located on the right bank of the Zenne River; (ii) heterogeneity of the streambed and aquifer and (iii) observation errors which all have margins of error.

The model was calibrated and validated. The optimal estimates of hydraulic conductivity for the layer just below the riverbed seem to be influenced by the different time steps used (hourly or daily) as a consequence of capturing different flow conditions. The obtained calibrated values for the whole parameter set, however, may still not represent one unique solution. It is possible to construct an equally well-calibrated model with different sets of parameters (the known 'equifinality' problem). Given the physically-based approach taken in this study, with parameters that are in principle obtainable by measurements, additional measuring campaigns can increase the level of certainty of the obtained parameters values. It should also be noted that obtaining a good match between the observed and simulated head and temperature data is necessary for obtaining a reliable estimate of flux and hydraulic conductivities. Calibrating the model with respect to observed temperatures only may compensate the errors in hydraulic head with errors in the hydraulic conductivity.

5 Groundwater-Surface water Interactions in Drained Riparian Wetland

5.1 Introduction

The riparian corridors separating a terrestrial ecosystem from a riverine ecosystem are areas of major groundwater-surface water interactions (Hayashi and Rosenberry, 2002; Triska et al., 1993). The word riparian was derived from the Latin word "*rip*" which means bank of a stream (Lowrance et al., 1985). Because of their strategic locations they generally exert significant control on water quality and quantity (Burt et al., 2010). The hydrology of near-stream riparian zones is complex mainly due to strong groundwater-surface water interactions such as inflow from hill slopes, upwelling from deeper strata, bank storage, overbank inundation etc (Schilling et al., 2006). According to Woessner (2000), groundwater-surface water exchanges and nutrient transport in the riparian zones is controlled by three main factors: (1) hydraulic conductivities distribution and magnitude in the riparian aquifer and within the channel; (2) the relation of stream stage to the adjacent aquifer water level; and (3) the geometry and location of the stream channel within the riparian plain.

The groundwater-surface water exchange, head gradient and hydraulic conductivity determine the residence time and the extent to which the riparian zone functions as a buffer (Burt et al., 2010) Lower residence time in the riparian zone reduces the effects of the buffer zone processes. In addition, lowering of the groundwater table in the riparian wetland causes aerobic conditions and hampers its effectiveness to reduce nitrate delivery to streams via denitrification (Burt et al., 2002; Hill, 1996; Schilling et al., 2006). Therefore, determining the magnitude and temporal variability of groundwater-surface water interaction is crucial for understanding nutrient dynamics in the riparian zones (Rassam et al., 2008; Schilling et al., 2006).

During flood events stream water levels rise in response to runoff, which results in water flow from the stream to the neighbouring aquifer. When the stream water level drops the infiltrated water is slowly released back to the stream (Whiting and Pomeranets, 1997; Woessner, 2000). These groundwater and surface water interaction phenomena, which occur during flood events are referred to as bank storage (Hunt, 1990; Whiting and Pomeranets, 1997). Bank storage provides the opportunity for transforming pollutants and nutrients (Peterjohn and Correll, 1984; Rassam et al., 2008).

Hence, the interaction between groundwater and surface water along riparian corridors have received increased attention (Burt et al., 2010; Sophocleous, 2002). Understanding the mechanisms of groundwater-surface water interaction further leads to better understanding of flow paths, residence time and the overall role of riparian zones in reducing nutrient loads within the wider landscape (Allan, 1995; Burt et al., 2010).

The Kielstau catchment in North Germany, federal state of Schleswig-Holstein, is situated in a lowland area and thus is characterized by flat topography, shallow groundwater level, low hydraulic gradient and flow velocities, and high interaction between groundwater and surface water (Fohrer and Schmalz, 2012). Investigation of nutrient transport and transformation processes in the riparian wetland of the Kielstau River, North Germany, has been conducted in several monitoring campaigns (Springer, 2006). Following this investigation a number of studies that characterize the water quality parameters and recorded water level both in the river and riparian wetland have been completed (Fohrer and Schmalz, 2012; Schmalz et al., 2008b). However, at present no study exists that reliably estimates groundwater-surface water flux exchanges in the riparian wetland. Therefore, the hydrologic connection between the riparian wetland and the river as well as the existing drainage ditches is poorly understood. This gap leaves the hydrologic connectivity between the riparian wetland and the river uncertain and hampers the understanding of the mobility and transformation of nutrients in the wetland.

This case is in fact a typical example when groundwater-surface water interactions need to be understood (and quantified) with limited available data. Modelling approaches are commonly brought in as support tools for realizing this task, but the quality of these models are often impaired by the same data limitations. Therefore, in order to increase the quality of the obtained results it is important to design and implement combined modelling approaches which involve the analysis and utilization of different available data sets. This is the main motivation in this study, where a detailed two-dimensional groundwater flow model was combined with an analytical profile model for one cross section of the wetland-river system. The two-dimensional model has been developed for the study area by using all existing available data combined with electrical resistivity tomography measurements, specifically conducted for this study. These data have been used for better estimation of the parameters of the underlying aquifer. This study presents how this combined approach can increase the quality of the obtained modelling results about the water flux exchanges in the riparian wetland. Such results can then also lead to better understanding of nutrient removal processes in the riparian zone. Overall, the study helps to establish the connection between groundwater, nutrient and land use and contribute to better understanding of nutrient removal and transport in future studies.

5.2   Study area

The study site is situated in the lowland catchment area Kielstau in the north of Schleswig-Holstein. Kielstau River flows along the southern border of the investigated riparian wetland area towards the west. This approximately 800 m long river section is about 4 m wide and about 1.5-2 m deep. The Kielstau River has been straightened at this section and exhibits a trapezoidal section.

The studied riparian wetland area is about 0.158 km^2 and is drained by ditches (Figure 5.1). The area is used as grassland. The 11 ditches in the study area that drain the riparian zone are located perpendicular to the Kielstau River. These ditches discharge water during winter period, but they fall dry during summer periods. As described in Schmalz et al. (2008b), the ditches are of different ages, type and length. Ditch one, constructed in 1999, has a length of 250 m and an average depth and width of 0.65 and 1.36 m respectively. Ditch four, constructed in 1990 has a length of 230 m and an average depth and width of 0.80 and 5 m respectively. Similarly, ditch eight, constructed and partly redirected in 1969/1970 has a length of 130 m and an average depth and width of 0.67 and 1.19 m respectively. Geologically the studied riparian wetland is characterized mostly as Sapric Histosols (Schmalz et al., 2009). According to Schmalz et al. (2009) the 1-2 m of the riparian soil which is characterized by medium to highly humified peat soil is underlain by carbonate rich peat clay. Whereas, the arable land in the North of the riparian wetland is characterized by sandy soil.

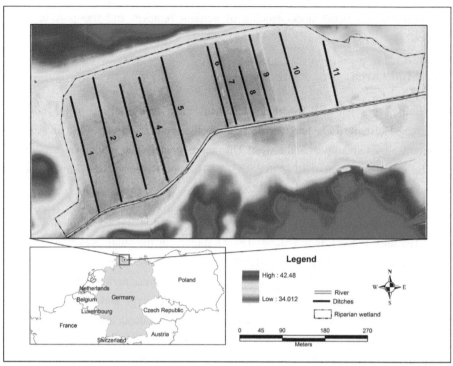

Figure 5.1: Digital elevation map of the study area, location of drainage ditches, and investigated riparian wetland along the Kielstau River

5.3 Methods

5.3.1 Electrical Resistivity Tomography

Electrical Resistivity Tomography (ERT) survey was carried out using multi-electrode resistivity meter (SYSCAL KID SWITCH 24) (Figure 5.2). The data were recorded using Dipole-Dipole sequence with 44 electrodes deployed along one profile line at an inter-electrode spacing of 5 m. The total length of the profile was 250 m. The measured resistivity values are apparent resistivities of a homogenous ground. To determine the true subsurface resistivity values, apparent resistivity values, obtained during data surveys, have been inverted using RES2DINV (LOKE, 2003). The noises due to the effect of electrodes were filtered during the inversion process. RES2DINV generates the inverted resistivity-depth image along the profile line. The quality of inversion result was checked by monitoring absolute error between the measured and predicted apparent resistivity. Resistivities of some common rocks and minerals are presented in Table 5.1. Inverted resistivity-depth is shown in Figure 5.3. Resistivity distribution of subsurface soil in these areas shows a significant variation of resistivity of soil at different depths along the profile line. The resistivity distribution at 1, 2.5, 4 and 6.5 m below the ground

88

surface respectively, varies from 16-796, 25-570, 26-300 and 21-720 Ω-m. The high local high resistivity values observed at each depth could be due to the presence of boulders.

Figure 5.2: Photo of ERT measurements along ditch one

Table 5.1: Resistivities of some common rocks and minerals (Source : Loke (2000))

Material	Resistivity (Ω-m)
Igneous and Metamorphic Rocks	
Granite	$5 \times 10^3 - 10^6$
Basalt	$10^3 - 10^6$
Slate	$6 \times 10^2 - 4 \times 10^7$
Marble	$10^2 - 2.5 \times 10^8$
Quartzite	$10^2 - 2 \times 10^8$
Sedimentary Rocks	
Sandstone	$8 - 4 \times 10^3$
Shale	$20 - 2 \times 10^3$
Limestone	50 - 400
Soils and waters	
Clay	1 - 100
Alluvium	10 - 800
Groundwater (fresh)	10 - 100
Sea water	0.2

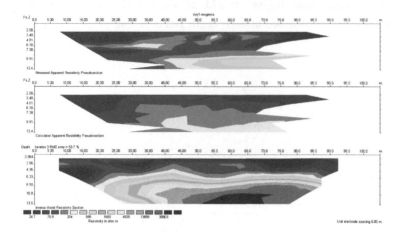

Figure 5.3: Measured apparent resistivity (Top), Calculated apparent resistivity (Middle) and Inverse Model resistivity (Bottom) values using RES2Dinv (starting point beginning of ditch one)

5.3.2   Water level data

Water level data used in this study were obtained from past studies in the study area (Schmalz et al., 2008b; Springer, 2006) . In the test site, various monitoring campaigns were completed in the past by the Department of Hydrology and Water resources management group of Kiel University since 2005 (Fohrer and Schmalz, 2012; Schmalz et al., 2008b). The location of groundwater observation wells and the location of the gauges in ditches are shown in Figure 5.4. Water level data monitored were collected at different temporal scales; ranging from hourly to weekly time scales. Ten groundwater observation wells were installed in the study area to study near surface groundwater dynamics. The wells were arranged in the form of transects along the drainage ditches (Figure 5.4). Additional observation wells were installed at the border of the riparian wetland (well 9 and well 10). Weekly groundwater levels were measured at an accuracy of 0.5 cm using an electric light gauge, while hourly readings were taken by pressure sensors (Schmalz et al., 2008a). Water levels of the Kielstau River were measured at the outlet of Ditch 1 (Gauge K) (Figure 5.4) using a pressure sensor.

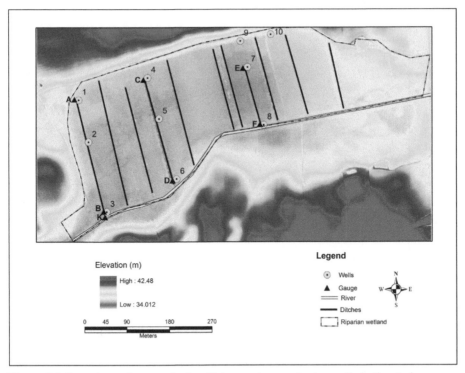

Figure 5.4: Locations of gauges in the ditches and groundwater wells in the study area

Water level in the ditches and the Kielstau River were measured at weekly intervals; additionally, hourly sampling was carried out at four selected measuring points (well 3, well 4 and Gauge K). Figure 5.5 shows the weekly observed water level for both Kielstau River and wells in the riparian wetland. As it can be seen from the figure, except well 5, the response of the water level in the observation wells is similar to the water level in the Kielstau River. Figure 5.6 also shows the weekly observed drainage ditches water level. Except Gauge A, all the water levels show similar response with the water level of Kielstau River. Gauge A shows more or less constant water level during the whole period. On the contrary, the water level in gauge D is higher than gauge C, may be due to adverse slope caused by siltation. Figure 5.7 shows the daily water level averaged from hourly observed water level in well 3, well 4, and Kielstau River. Temporal variability of daily water level data is more obvious than that of the weekly water level data, which shows that more details of the water level dynamics is captured by higher temporal resolution data ranging from daily to sub-daily measurements. In contrast to water level data, nitrate concentrations were only measured (weekly) at 3 locations, which is insufficient for modelling studies.

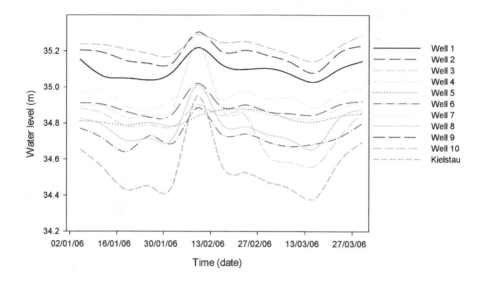

Figure 5.5: Weekly measured water level above mean sea level in observation wells spatially distributed in the riparian wetland and in the Kielstau River

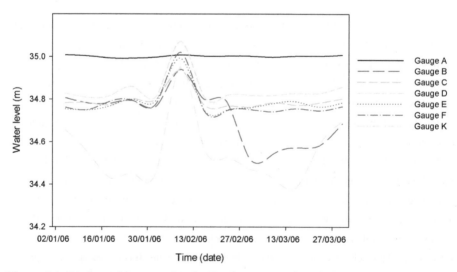

Figure 5.6: Ditch weekly water levels (for Gauges A- F) and river water level (Gauge K)

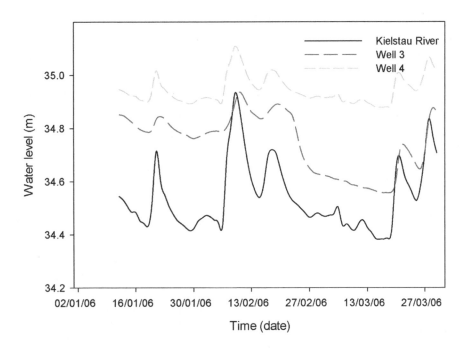

Figure 5.7: Daily water levels averaged from hourly measurements (for well 3 and 4) and river water level (Gauge K) for the study period

5.3.3   Numerical Groundwater Flow Model

The groundwater flow system was modelled using the MODFLOW code (Harbaugh et al., 2000) in Processing MODFLOW modelling environment (Chiang and Kinzelbach, 1998). The model domain covers an area of 0.158 km^2. The detailed 1m x 1m Digital Elevation Map (DEM) obtained from LVA (2008) was re-sampled into 4 m x 4 m grid so as to allow detailed assessment of riparian groundwater conditions and groundwater-surface water exchange at reasonable computation time. Land surface elevations (for the top layer) were then extracted from the re-sampled 4 m DEM. The finite-difference mesh consisting of 111 rows and 209 columns was constructed with a uniform grid size of 4 m. In the vertical directions, the model consists of one layer with constant layer thickness of 7 m. This was decided on the basis of the electrical resistivity tomography data, which indicate that on average below this depth the aquifer material is predominantly with very low permeability. For temporal discretisation, the three month simulation period (January-March 2006) was divided in 12 stress periods where each stress period corresponds to one week during which all inputs are constants.

Boundary conditions

Along each of the boundaries of the aquifer, different types of boundary conditions were specified. Along the northern side of the model domain, a specified head boundary conditions, based on the observed ground water level data (well 1, well 9 and well 10) was used. In the southern side, Kielstau River water level was used as boundary. In the eastern and western boundary, a no flow boundary was assumed. Figure 5.8 shows the boundary conditions used in the study area. In the vertical direction, the model is bounded by a recharge flux at the top and by impermeable soil layer at the bottom.

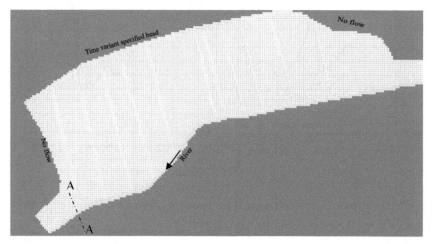

Figure 5.8: Model boundaries

River/Drain -Aquifer interactions

The flow between a river and an aquifer depends on the difference between the water level in the river and the groundwater head. Multiple MODFLOW packages have been developed in the past to represent the interaction between groundwater and distinct types of water bodies. The MODFLOW river and drain packages were used in this study to model the river and aquifer, and drain and aquifer interaction, respectively.

The river package was used to simulate the river boundary condition in the riparian groundwater model. It allows the river cell to be a sink if flow is toward the river (gaining stream) or a source if flow is out of the river (losing stream). River fluxes are calculated by the model using a riverbed conductance term and the driving head difference between the river stage and the calculated groundwater elevation in the cell. The river package is used to simulate head-dependent flux boundaries. The riverbed elevation, river stage, and conductance of the riverbed material need to be specified. Riverbed elevations are available along the reach at some points. These points were

94

linearly interpolated along the entire reach to obtain the riverbed elevation for every grid cell. The riverbed bottom elevations were taken as one meter below the actual riverbed elevation. The conductance term incorporates assumptions regarding vertical hydraulic conductivity, bed thickness and wetted area with the model cell. This lumped parameter was estimated using model calibration. River stages were determined from observed stage at gauge K located at the downstream end of ditch one.

The MODFLOW drain package was used to simulate the effect of the drainage ditches on the groundwater flow regime. MODFLOW drain package (Harbaugh et al., 2000) was developed to simulate agricultural tile drains that remove water from an aquifer at a rate proportional to the difference in water level between the aquifer and some fixed drain elevation as long as the head in the aquifer is above that elevation. If the water level in the aquifer falls below that of the drain, no additional water removal occurs. In other words, the drainage level is assumed to be equal to the highest level that the groundwater will reach before it is discharged. The drain elevations were fixed based on the measured water level in the drains. For the ditches with no measurements drain elevations were interpolated linearly from those available drain elevations. In addition to a drain elevation, the drain conductance needs to be specified in the drain package. This lumped parameter incorporates information on characteristics of the drain and its immediate surroundings, as well as the head loss between the drain and the aquifer. Drain conductances were adjusted during the calibration process.

Hydraulic conductivity estimation

Hydraulic conductivity is one of the highly variable parameter in space. In most of the cases our knowledge regarding the values of hydraulic conductivity parameter is limited to randomly located point measurements or no measurement at all. The standard techniques to determine hydraulic conductivity, such as pumping test, grain size analyses are expensive. On the other hand, the use of geophysical methods such as electrical resistivity tomography provides complementary information and might help to reduce the cost of hydrogeological investigation (Berryman et al., 2000; Loke, 2000). They also provide much superior spatial data and are not invasive.

The following procedure was used to compute the hydraulic conductivity of the aquifer from the inverted resistivity values obtained from ERT survey.

    i.   In a fully saturated, clay-free granular formation, the Formation factor (F), originally defined by Archie (1942) is calculated by dividing the ground resistivity values (ρ_o) obtained from inversion to water electrical resistivity

ii.   values (ρ_w) $(F = \frac{\rho_o}{\rho_w})$. Resistivity of water was assumed from literature (porewater resistivity assumed from EC measurement at ditch one is 20 Ω-m)

ii.   According to Archie's (1942) the formation factor can be expressed as $F = a *$ $\emptyset^{-m}$. Usually the value of a = 1, and the cementation degree varies between 1.3 and 2.5. Assuming m=1.5 (low cementation degree) the porosity values for each depth have been calculated.

iii.   Assuming equivalent grain diameter equal to 0.03 mm specific surface S for spheres with diameter D is calculated as $S = \frac{6}{D}$

iv.   Hydraulic conductivity K (m/s) for different grain sizes at various porosities could be then computed using the modified form of Kozeny-Carmen equation $K = \frac{g}{\varepsilon}\frac{c}{S^2}\frac{\emptyset^3}{(1-\emptyset)^2}$ (Bear, 1988; Urish, 1981). Where: g is acceleration due to gravity (m/s^2); ε is kinematic viscosity of porewater (m^2/s); c is shape-tortuousity factor (for spherical particles c=1/5), S is specific surface for spheres with diameter D (m).

In the present study the attempt was to estimate average hydraulic conductivity from the resistivity values observed in the field. The above methods that relate porosity, formation factor and grain size distribution were used to determine the hydraulic conductivity. As it can be seen in Figure 5.3 the average resistivity for the depths in between 0 - 6 m can be assumed to be 80 Ω-m. Further, assuming the resistivity of pore water 20 Ω-m and degree of cementation and grain size of 1.5 and 0.3mm respectively, the computed hydraulic conductivity could have a value of 1 m/d. If we consider the highest resistivity observed in the field which is 720 Ω-m, the computed hydraulic conductivity would lower to a value of 0.001 m/d. Previous field studies also reported the same order of hydraulic conductivity values in the study area, ranging between 2.94 m/d and 0.15 m/d. Moreover, the computed porosity values were in the range of field measured porosity values. Therefore the initial estimate of the hydraulic conductivity for the model (constant for the whole modelling domain) was taken to be 1 m/d. This value was later adjusted during the calibration process.

Recharge estimation

Groundwater recharge estimation is one of the key components in any groundwater flow and contaminant transport model. Various methods for recharge estimation exist in literature (e.g. direct measurements, water balance methods, Darcian approach, tracer techniques, and empirical methods) (De Vries and Simmers, 2002). The subject of choosing appropriate techniques for quantifying groundwater recharge is addressed by Scanlon et al. (2002). Like many typical unconfined aquifers, the primary input to the aquifer in the study area is considered to be groundwater recharge from precipitation.

Since groundwater recharge vary with time and space these variations are difficult, to accurately estimate. Many modelling studies obtain recharge rates through model calibrations (e.g. Schilling et al. (2006)). For this study, recharge estimates were obtained from previous studies in the area that used SWAT model (Soil and Water Assessment Tool) (Arnold and Soil, 1994) for the Kielstau watershed (Schmalz et al., 2008c). Figure 5.9 shows the recharge values obtained from the available SWAT model developed for the area.

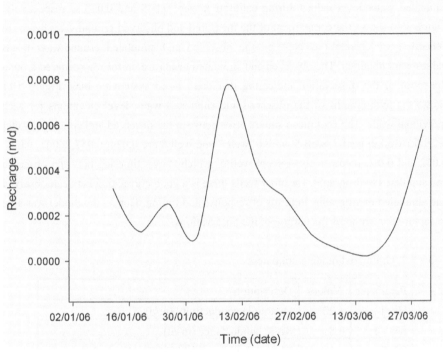

Figure 5.9: Recharge estimates obtained from SWAT model.

Model calibration

The groundwater flow model was calibrated using observed heads from seven monitoring wells measured from January 2006 till February 2006. The calibration was performed using the automated parameter estimation code PEST (Doherty, 2008) and Genetic Algorithm Toolbox in MATLAB (Chipperfield and Fleming, 1995). Steady state model was first calibrated so as to create steady state head solutions for initial conditions for transient model calibration.

The transient model was set up with a weekly stress period for the study period for which observation data were available. Specific yield for the aquifer was estimated using laboratory tested soil samples collected at 1-1.5 m depth. A specific yield of 0.14 was applied uniformly to the entire model. The calibration was carried out based on 84 groundwater heads measured in seven observation wells. The hydraulic conductivity values, river conductance and drain conductance found during model calibration are presented in Table 5.2. The calibrated hydraulic conductivity is within the range of hydraulic conductivity values determined using the geophysical method.

The root mean square error and mean absolute error between the observed and simulated water level values during calibration were 0.049 and 0.034 m respectively. Figure 5.10 shows the comparison of the measured and observed groundwater heads. In general, good correlations between the observed and simulated groundwater heads values were obtained. The observed and simulated heads are uniformly scattered around the mean of the observation indicating that there is no systematic bias. Figure 5.11 shows the hydrographs of the observed and simulated water level elevations for each individual wells. The root mean square error between the observed and simulated water level for well 2, well 3, well 5, well 6, well 7 and well 8 are 0.034, 0.057, 0.041, 0.027, 0.052 and 0.053 respectively. Except well 5, which shows time lag between observed and simulated hydrographs, for other wells there is a good correlation between observed and simulated hydrographs in terms of response and timing. Hence, the model was able to capture the temporal fluctuation of the observations.

Table 5.2: Optimized model parameters

Parameter	Values	Description
Hk_1	1.22	Hydraulic conductivity (m/d)
RIV_1	0.81	River conductance (m^2/d)
DRN_1	6.38	Drain conductance (m^2/d) for drain one
DRN_2	6.07	Drain conductance (m^2/d) for drain two and three
DRN_4	9.07	Drain conductance (m^2/d) for drain four
DRN_91	0.62	Drain conductance (m^2/d) for drain nine and ten
DRN_8	6.29	Drain conductance (m^2/d) for drain eight
DRN_67	3.32	Drain conductance (m^2/d) for drain six and seven
DRN_5	7.03	Drain conductance (m^2/d) for drain five
DRN_11	3.24	Drain conductance (m^2/d) for drain 11
SS	6.62E-4	Specific storage
Sy	0.14	Specific yield (measured)
n	0.15	Porosity (measured)

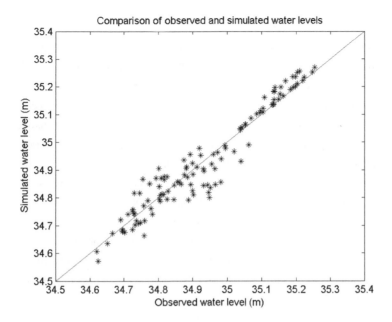

Figure 5.10: Observed and simulated scatter water levels for steady state model calibration.

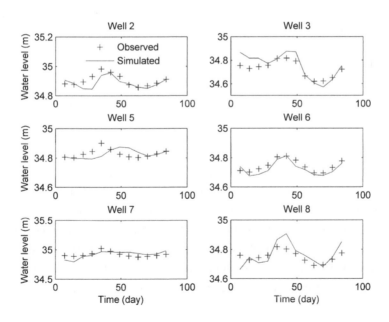

Figure 5.11: Observed and simulated hydrographs at monitoring wells for transient model calibration.

5.3.4 Analytical Model

Analytical models offer an inexpensive way to estimate groundwater-surface water interactions (Spanoudaki et al., 2010). In this study a computer program STWT1 (Barlow and Moench, 1998; Desimone et al., 1999; Moench and Barlow, 2000) was used to estimate groundwater discharge rates in response to the variation in recharge and stream stage. The program uses the convolution method to solve the analytical solution for time-varying stream-stage or recharge inputs (Barlow et al., 2000). A number of simplifying assumptions regarding the groundwater system were considered to obtain the analytical solution. Details about the governing equations, solution methods and assumptions can be found in Barlow and Moench (1998). In this study the analytical profile model was applied at cross-section A-A (Figure 5.8) and the groundwater flow was assumed to be two dimensional in the vertical plane perpendicular to the stream with semi-impervious stream bank material (Figure 5.12). The stream is considered to be, fully penetrating the saturated thickness of the aquifer which is assumed to be 7m.

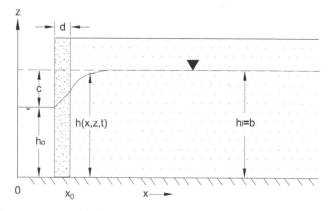

Figure 5.12: Semi-infinite water table aquifer with semi-pervious stream-bank material (b, saturated thickness of aquifer; c, instantaneous step change in water level of stream; d, width of semi-pervious stream-bank material; h(x, z, t) head in the aquifer as a function of distance from the middle of the stream (x), vertical coordinate (z), and time (t); hi, initial head and stream stage; ho, water level in stream after step change; xo distance from the middle of stream to stream-aquifer boundary) Source (Barlow & Moench, 1998)

Analytical model calibration

Various hydraulic properties of aquifer and semi-pervious stream-bank material affect groundwater level and seepage rates. For the water table aquifer, the case for this study, the relevant properties are vertical and horizontal hydraulic conductivity, ratio of vertical and horizontal hydraulic conductivity, specific storage, and specific yield and hydraulic properties of the stream-bank material. The hydraulic properties of the stream-bank material is lumped to one parameter termed retardation coefficient (Hall and Moench, 1972) or stream bank leakance term (Barlow and Moench, 1998). According to Hall and Moench (1972) and Barlow and Moench (1998), retardation coefficient or stream-bank leakance term (a) can be related to: the width of the semi-pervious stream-bank material, horizontal hydraulic conductivity (K_x), and hydraulic conductivity of the stream-bank material in the direction perpendicular to stream flow (Lappala et al.). This is described by the relation: $a = \frac{K_x d}{K_s}$. Higher value of leakance can result from either increased stream bank thickness or decreased stream bank hydraulic conductivity relative to the hydraulic conductivity of the aquifer. Thus, higher leakance values represent increased hindrance by the stream bank material to the transfer of water across the stream-aquifer boundary, which consequently results in reduced groundwater levels, seepage rates, and bank storage volumes in the aquifer (Desimone et al., 1999).

Hourly stream stage and groundwater level were averaged to obtain daily values needed for model calibration. Observation well close to the river (well-3) was used for calibration. The calibration period was from January-February, 2006. This is because: ditch one was maintained in February, 26, 2006 and during this period the channel geometry was modified from the centre till the end of the drainage ditch. This modification consequently altered the conductance value. Hence the observed water level in the well 3, which was used as a calibration point, is unusually lower than the preceding period. This is because the groundwater flow is directed to the ditch rather than to the river. These phenomena cannot be handled by the analytical model due to the assumption of homogenous medium and time invariant aquifer parameters. Therefore, unlike the numerical model, the analytical model was calibrated for only two months considering the period before February, 26, 2006.

As part of the calibration process the hydraulic properties of the aquifer and semi-impervious streambed material were varied. Figure 5.13 shows the observed and simulated groundwater levels at well 3. As it can be seen, good agreement was obtained during the calibration period. The calibrated values were presented in Table 5.3. The root mean square error between the observed and simulated water levels is 0.029. The hydraulic conductivity obtained with this method is lower than the numerical groundwater model. It should be noted that Barlow et al. (2000) demonstrated that, without independent knowledge of stream-bank seepage rate or either hydraulic conductivity or specific storage it would not be possible to determine a unique value of hydraulic conductivity or specific storage of the aquifer during calibration, rather it is only possible to determine an effective diffusivity for the aquifer. Aquifer diffusivity is the ratio of the aquifer transmissivity (T) to aquifer storage coefficient (S) (Barlow and Moench, 1998; Hall and Moench, 1972).

Table 5.3: Calibrated analytical model parameters

Paramterer	Values	Describtion
Kh	0.47	Horizontal hydraulic conductivity (m/d)
ASY	0.14	Aquifer specific yield
XKD	0.12	Ratio of vertical to horizontal hydraulic condctivity
XAA	2.5	Streambank leakance (m)
AS	0.001	Specific storage (m^{-1})

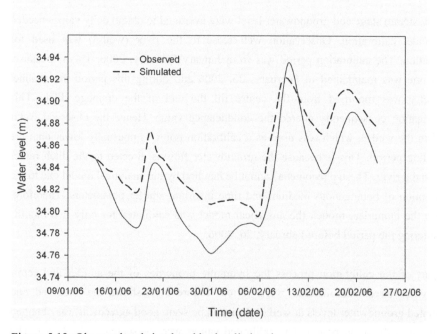

Figure 5.13: Observed and simulated hydraulic heads.

5.4   Results and Discussions

Numerical groundwater flow model

In this section, the simulation results obtained from the numerical groundwater flow model is presented. The water table contours simulated with the numerical groundwater flow model corresponding to stress period four is shown in Figure 5.14. For ease of comparison with the analytical model the river-aquifer flux exchange obtained from the numerical groundwater model was presented in terms of discharge per unit width as shown in Figure 5.15. The river-aquifer fluxes were computed using the sub-regional water budget at the river grid cell located close to well3. The total water budget of the model domain simulated with the numerical groundwater flow model consists of: recharge 204 m^3/d, inflow from the specified head cells of 34.3 m^3/d, outflow through river leakage of 23.6 m^3/d, outflow through the drains 132.5 m^3/d and subsurface storage of 82 m^3/d.

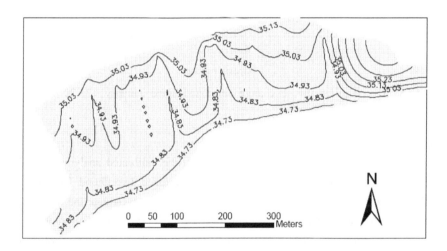

Figure 5.14: Groundwater contour maps corresponding to the fourth stress period (high exchange from aquifer to river).

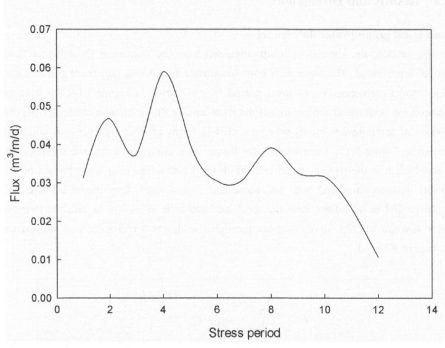

Figure 5.15: River-aquifer flux exchange simulated since January 12, 2006, using MODFLOW.

Analytical model

The river-aquifer flux exchange was computed with the calibrated analytical profile model. Monte Carlo simulations based on the Generalized Likelihood Uncertainty Estimation (GLUE) (Beven and Binley, 1992) was used to investigate the parameter uncertainty of the calibrated model parameters. Thus, 10000 simulation runs were carried out using parameter ranges shown in Table 5.4. Model runs giving root mean square error higher than 0.05 have been rejected, while the others are considered as behavioural models. The threshold for the rejection criteria was set based on the water level measurement accuracy. From the ensemble simulations, the 5th, 50th and 95th percentiles were computed (Figure 5.16).

Table 5.4: parameter ranges used in the Monte carol simulations

Paramterer	Minimum	Maximum	Describtions
Kh	0.25	5	Horizontal hydrulic conductivity (m/d)
ASY	0.1	0.25	Aquifer specifc yeild
XKD	0.12	1	Ratio of vertical to horizontal hydrauilc condtivity
XAA	0.001	50	Streambnk leakance (m)
AS	0.0001	0.025	Specific storage (m^{-1})

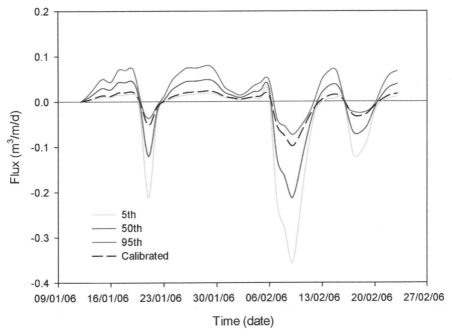

Figure 5.16: Simulated flux rates model parameter uncertainty (5th, 50th and 95th percentiles), positive values represent groundwater discharge and negative values indicates groundwater infiltration into the aquifer.

Comparison of groundwater-surface water fluxes exchange estimates

Results of the study presented in Figure 5.17 showed that the fluxes between the river and aquifer computed with the numerical groundwater model are all positive indicating that the river is gaining fluxes from the aquifer. On other hand, the analytical model correctly simulated the reversal of groundwater flow indicated by observed water levels. The positive fluxes computed with the two models matches very well. The

105

discrepancies between the models estimates can be attributed to different factors: (i) temporal resolutions of water levels used to calibrate the models; weekly measured water level boundaries likely dampen, the magnitude of fluxes during high rainfall events (ii) averaging the daily fluxes estimates to obtain weekly values. Ebrahim et al. (2012) showed that low resolution temporal water level data is unable to capture the fluxes exchange that may occur from the river to the aquifer during short flood events as a result of the averaging effect of larger time interval, and (iii) the assumptions that are inherent to the models.

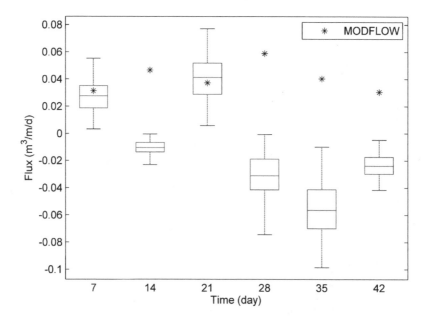

Figure 5.17: Box plot of river-aquifer flux exchange simulated with analytical model and flux estimates using the calibrated MODFLOW model

The analytical model in this study was applied at one location normal to the river channel which is close to monitoring well 3. Thus, the river-aquifer fluxes exchange estimates at this particular point was compared with numerical model. Such comparison is important to gain better understanding of the river-aquifer exchange at different temporal scales. Spatial variability in the hydraulic conductivity of the aquifer and riverbed sediments results in spatial variations in rate of river-aquifer exchange. Especially, when aquifer and riverbed heterogeneity is high single point comparison may not be sufficient.

5.5 Conclusion

Transient groundwater model was set up for the Kielstau riparian wetland for the period of three months (January - March) year 2006. Results of the numerical groundwater model shows that the outflow through the drainage ditches accounts for 64% of the total recharge into the riparian wetland while outflow through the riverbed accounts for 12% of the total recharge into the riparian wetland. The drainage ditches network is intended for lowering the water table in the riparian wetland, however, they also alter significantly the groundwater flow paths. Variation in groundwater discharge to changes in stream stage were investigated using the transient groundwater model and analytical profile model at the river-aquifer interface close to observation well 3. Results show that reversal of flow from the stream to the aquifer is only captured by the analytical profile model. Although, the analytical model could not incorporate all the complexities of the real aquifer system in the study area it provided an insight for the rate of flow from the river to the aquifer, which was not obtained with the numerical model, due to limited availability of data at higher resolution. Detailed knowledge of the river-aquifer interaction may have significant implications for further studies related to nutrient transport in this interface, which suggests the need of daily to sub-daily resolution monitoring data.

6 A Simulation-Optimisation Approach for Shallow Groundwater Level Management in Drained Riparian Wetland

6.1 Introduction

The role played by riparian zones in water quality improvement highly influenced by the hydrology of the riparian zone (Burt et al., 2010). The groundwater flow through the riparian zone should provide sufficient residence time for nutrient removal and it should also be high enough to reach the plant root zone so that denitrification process to occur (Hill, 1996). This is because denitrification process is more effective in the plant root zone where sufficient organic material is available for denitrifying bacteria (Osmond et al., 2002). Other studies also showed that denitrification in agricultural soils mainly occur under saturated or near saturated conditions (Evans et al., 1995; Thomas et al., 1992). Therefore, for effective removal of nutrient especially nitrate in the shallow groundwater, the groundwater that is flowing through the riparian zone must pass through the zone where plant root can absorb the nitrate or provide organic matter for microbes to denitrify the nitrate (Osmond et al., 2002). However, if some of the groundwater passes the riparian zone fast through drainages systems or at a depth where root of a plants in the riparian zones have no interaction with the groundwater the riparian zone will have negligible contribution in nitrate removal (Stone et al., 1995; Stone et al., 1992; Thomas et al., 1992).

Drainage systems have been used to remove excess water from water logged agricultural soils. Without proper drainage systems, soils with poor natural drainage will remain waterlogged for several days after heavy rainfall events (Osmond et al., 2002). According to Evans et al. (1991) agricultural drainage systems are implemented to offer two main advantages: (1) to increase crop yield on poorly drained soils in particular during wet season, and (2) to improve field conditions such as trafficability during tillage, seed bed preparation, planting, and harvesting. On the other hand, agricultural drainages are of real concern to the environment (Evans et al., 1991; Strock et al., 2010). One of the main concerns is water quality impact on receiving water bodies. If they are not properly designed and managed, they increase loss of nitrate to receiving water bodies.

To decrease nitrate losses from drainage systems, water level management through controlled drainage was found to be successful (Evans et al., 1989, 1991; Skaggs and

Gilliam, 1981). The idea of controlled drainage practice is to place water control structures at various location of the drainage system so that the water level in the drainage system is raised. This practice is found to be economical because as the slope decrease less water control structures are needed. Effective water quality improvement through controlled drainage system can be obtained only when the water level in the drainage ditches is managed at a correct water level (Osmond et al., 2002).

The drained riparian wetland presented in chapter 5 is a typical case where we believe controlled drainage systems need to be practiced. It is important to control the nitrate export from the drained riparian wetland by managing the drainage systems such that only minimum drainage water necessary is allowed to exit the riparian zone. In similar hydrological stetting North-Eastern Germany, Tiemeyer et al. (2007) observed that denitrification in groundwater are higher in areas were the groundwater levels are high. And they recommended using controlled drainage systems so that to raise the water level in the ditches to ensure low hydraulic gradients. Tiemeyer et al. (2007) reported that lowland soils mainly composed of peat are not passable with machinery in winter period. Drainage systems, necessary to provide trafficability during wet periods need to be designed appropriately so as to allow minimum drain outflows.

In this study a simulation-optimisation approach was used for designing appropriate water level elevations for the drainage ditches found in the drained riparian wetland described in chapter 5. The use of simulation-optimisation approach for groundwater management problems such as aquifer remediation, conjunctive use of groundwater-surface water etc are well established (see for an overview, although not very recent Gorelick (1983) and Wagner (1995)). However, there have not been such studies for groundwater level management in the drainage systems particularly for drained riparian wetlands. The main advantage of the simulation-optimisation approach is that the dynamics of the physical system, represented by the simulation model, can be incorporated in optimisation procedures designed to minimize or maximize certain management objective.

6.2 Methods

6.2.1 Simulation Model

A transient, two dimensional groundwater flow model developed in chapter 5 was used in this study. As described in chapter 5, in the riparian wetland there are 11 drainage ditches that remove excess overland flow and subsurface flow from the riparian zone.

Shallow groundwater flows into the drains were simulated using MODFLOW drain package. Water levels measured in the drainage ditches were provided as input to the drain package, so that the drain outflow based on the difference in water level between the adjacent aquifer and water level in the drains can be simulated. The transient groundwater model was calibrated for the period of January 2006 till March 2006 assuming weekly stress periods. The three month period was divided in 12 stress period where each stress period corresponds to one week during which all inputs are constant. For the purpose of this study stress period, which correspond to high ground water level in the drained riparian wetland was selected to evaluate the water level management objective.

6.2.2   Formulation of optimisation problem

The objective of the management problem defined in this study was to minimize the total drain outflow. Minimizing the total drain outflow was used as a surrogate for reducing nitrate outflow from the shallow groundwater. Although the objective function is linear, the optimisation is non-linear because the changes in water level in the drains do not respond linearly to changes in head in the unconfined aquifer. The non-linear optimisation problem was then solved using global optimisation tools combined with the simulation model based on MODFLOW, which relates the state and decision variables.

Water levels in each of the existing drainage ditches were used as decision variables. Upper and lower limits of water levels in each drain were also placed as additional constraints.

$$\text{Objective function} = \text{Minimize} \sum Qi \qquad\qquad (6.1)$$

Subject to: water level constraints at each control point in the model domain
Decision variables: drainage ditches water level
Where: Qi is the drain outflow rate of ditch i

The constraint limit was assumed based on Evans et al. (1991) studies. The authors prepared a controlled drainage guideline that can help to achieve both water quality and crop production benefit for different crops at different growing season. For fallow fields they recommended groundwater level to be 30-45 cm below the land surface so as to minimize drainage outflow and encourage denitrification. In this study to allow trafficability it was assumed that lowering the groundwater level 25 cm below the land surface is enough. These constraints were specified in the model domain at every five grids (at 20 m spacing's).

For optimal estimation of the drain water levels, two global optimisation techniques; GA and CRS4, were coupled with MODFLOW using GLOBE.

6.3   Results and Discussions

The performance of the shallow groundwater management system design was evaluated in terms of total drain outflow and its ability to satisfy trafficability. Simulations were conducted for various drainage levels and the results were analyzed in terms of violation of constraints - the ability of the drainage systems to satisfy trafficability requirements during the wet periods. The outflow and percentage of constraint violations is presented in Table 6.1. Using simulation model alone in order to satisfy trafficability in the riparian zone during wet period the drain elevations need to be lowered by 1 m from the current conditions. This means a total drain outflow of 1000 m^3/d need to be removed. The water budget result show that the existing total drain outflow is about 187 m^3/d which is far less than the trafficability requirements.

On the hand, the simulation-optimisation results showed that only 2/3 of the total drain outflow computed by simulation model is necessary to be removed. Furthermore, the results of the two global optimisation algorithms used in the study showed good agreement. The total drain outflow by CRS4 and GA were 642 m^3/d and 663 m^3/d, respectively. Though, CRS4 provides most accurate solution the number of model runs compared to GA is slightly higher. CRS4 finds the optimal solution after 7433 model runs while GA achieve its optimal solution at 7184 model run. The optimal drain elevations obtained with the two algorithms is presented in Table 6.2. In average the drain elevations need to be lowered by 0.6 m with respect to the existing drain elevations.

Table 6.1: Scenario analysis results obtained using simulation model

Scenarios	Depths to which Water levels in the existing drains to be lowered (m)	Total drain outflows (m^3/d)	Percentage of constraint violations (%)
1	0.2	360	21
2	0.4	543	7.4
3	0.5	636	4.5
4	0.6	729	2.5
5	0.7	822	2
6	0.8	915	1
7	1	1099	0.5

Table 6.2: Optimal drain water levels obtained using simulation-optimisation

Descriptions	Conductance (m^2/d)	Existing Water level (m)	Optimal water level computed using GA (m)	Optimal water level computed using CRS4 (m)
DRN-1	6.38	34.997	34.65	34.52
DRN-2	6.07	34.78	34.27	34.44
DRN-3	6.07	34.78	34.56	34.56
DRN-4	9.07	34.96	34.22	34.27
DRN-5	7.03	34.77	34.23	34.32
DRN-6	3.32	34.75	32.88	32.82
DRN-7	3.32	34.75	34.15	34.19
DRN-8	6.29	34.999	34.52	34.54
DRN-9	0.62	34.73	33.73	33.81
DRN-10	0.62	34.73	34.72	34.70
DRN-11	3.24	34.73	34.72	34.73

6.4 Conclusions

Simulation-optimisation approach was used to predict the total drain outflows from artificially drained riparian wetland with high water tables. Total drain outflows were determined for alternative drainage system designs and operational procedures. The results showed that trafficability can be satisfied by several different drainage systems designs. For the drained riparian wetland considered in this study, there was a three-fold difference in drainage outflow between existing practice and drainage system design that satisfy drainage objectives. The amount of drain outflow that leaves the riparian wetland can be reduced by using controlled drainage during drier period. Otherwise, drainage systems designed to remove excess water during wet period may lead over drainage during drier periods.

The simulation-optimisation modelling approach calculates better solutions than trial-and-error simulation modelling alone. In this study we focused on drain elevation optimisation however, similar results can be achieved by optimizing the spacing between the drains.

7 A Simulation-Optimisation Approach for Evaluating Feasibility of Managed Aquifer Recharge

7.1 Introduction

Rapid increase in population and global climate change may result in shortage of long-term supply of freshwater in many parts of the world (Maliva and Missimer, 2010). In arid countries where surface water resources are mostly absent and groundwater resources are often non-renewable, the issue of water shortage is more likely (Al-Rashed and Sherif, 2000; Mays, 2013). During the last decades, the use of groundwater resources has become particularly intensive in coastal areas due to the increase in urbanization, touristic development, and irrigated land expansion (Singh et al., 2013; Voudouris, 2011). As a result, negative water balance is established in coastal aquifer systems triggering seawater intrusion that has negative consequences for the socioeconomic development of these areas. Under scarcity conditions, withdrawals during the dry seasons often by far exceed the safe yield of aquifers, resulting in the depletion of reserves, allowing water of inferior quality to intrude into the aquifers (Pereira et al., 2002). For example, Abdalla and Al-Rawahi (2012) reported that, since 1970s groundwater quality in the Oman Coast has deteriorated as a results of progressive advancement of saline plume. Managed aquifer recharge, also referred as artificial recharge is widely practiced in many countries to enhance water supplies particularity in arid and semi-arid regions.

Interest in artificial groundwater recharge is growing due to many reasons such as preventing the decline of groundwater levels, preventing seawater intrusion, securing and enhancing water supply, and maintaining environmental flows (Asano and Cotruvo, 2004; Pyne, 1995). A number of different techniques have been developed to recharge the aquifer. The conventional methods include surface infiltration and injection wells (Bouwer, 2002). Aquifer storage and recovery (ASR), a term coined to aquifer recharge via wells is one of the methods that is applied widely as a means of storing excess water in the aquifer during periods when seasonal demand is less than available supply, with the intention of later withdrawal for use when the demand is high (Maliva and Missimer, 2010; Pyne, 1995). This has certain advantages over surface storage in tanks or reservoirs, such as reduced land use, decreased risk of contamination, and reduced loss to evaporation. With regard to recovery of the injected water two mechanisms exists: (1) injection of water into a well for storage and recovery from the same well, referred as aquifer storage recovery (ASR) and (2) injecting water into a well for storage, and recovery from a different well, known as Aquifer Storage, Transfer and Recovery. In most instances, injection and recovery from the same well is the preferred option for economic reasons. It is typically less expensive to construct one dual purpose well than dedicated injection and recovery wells. However, using separate injection and recovery wells may also be desirable to improve stored water quality by providing additional residence time and to take advantage of aquifer filtration.

Due to growing scarcity of available land and substantial losses of water from surface reservoirs by evaporation, recharge using surface storage may not be feasible in arid countries like Oman. Therefore, subsurface storage using injection wells is chosen as an alternative. Freshwater resources in Oman are scarce. Conversely due to increase in connection rates in the capital of Muscat, there is an increase in wastewater production, which can be further treated and used for higher value use compared to the current use of irrigation of public parks (Zekri et al., 2013). In the Muscat region of Oman, the Oman Wastewater Services Company called HAYA is mandated to collect, treat and dispose domestic wastewater. According to Ahmed et al. (2013) HAYA anticipates the percentage of households in the sewer network to increase to 80% in 2018 and a surplus of up to 100,000 m^3/day of treated wastewater (using Membrane bio reactor) would become available during winter months, after meeting the demand for irrigating city parks, roadside plants and trees. Therefore, the source of water for MAR is considered to be tertiary treated wastewater (to the level of potable water quality), which many not need additional water treatment. ASTR method was selected for this study mainly for appropriate localization of the recovery wells at the point of use and for safeguarding the psychological barrier that prevents people from using wastewater. Injecting treated wastewater into vulnerable coastal aquifer regimes has many benefits such as restoration of groundwater levels, prevention of seawater intrusion, and balancing seasonal differences in water demand (Asano and Cotruvo, 2004).

The capacity of the aquifer to store water is one of the most critical factors in selecting a site for underground storage system. Once potential sites are identified, hydrological evaluations should be carried out in order to check the capacity of the aquifer to fully assimilate the available recharge flows. Insufficient capacity may result in groundwater mounding, lateral movement of water to discharge points, and reduction of the unsaturated zone thickness (Pyne, 1995). The selection of an appropriate storage zone is an important consideration, impacting costs, the ability to get water in and out of the storage zone, and the potential for water quality impacts. As will be shown, many of these design aspects are addressed by the simulation-optimisation methodology presented in this study, however, without, explicitly addressing the water quality impacts.

The purpose of this study is to assess the feasibility of managed aquifer recharge and to present planning level model for optimal development and operation of aquifer recharge and recovery in the Samail Lower Catchment, Oman. The questions to be answered are: (1) how much is the assimilative capacity of the aquifer without causing undesirable groundwater mounding? (2) how should the injection wells rate be proportioned so as to achieve the maximum total injection? (3) where should we locate the recovery wells and what should be their pumping rates in order to maximize the recovery percentage of the recharged water without causing further seawater intrusion into the coastal aquifer? (4) would such recovery wells configuration (location, pumping rates) allow for sufficient travel time of injected water, so that water quality cannot be hampered? (5) How much is the recovery efficiency?

7.2   Materials and Methods

This study focuses on the application of simulation-optimisation model for studying the feasibility of artificial recharge or managed aquifer recharge. Successful planning and management of artificial schemes often require consideration of different groundwater management objective and set constraints. Using simulation-optimisation approaches an optimal aquifer recharge schemes that meets a given objective and set of constraints can be identified. However, little work has been published on simulation-optimisation of artificial recharge systems. Moreover, the simulation-optimisation models developed in the past for artificial aquifer recharge studies in the unconfined aquifer system assumed the unconfined aquifer problem as linear problem though the system behaves nonlinearly in nature.

The simulation-optimisation models developed for unconfined aquifer in the past studies were mostly based on groundwater simulation model, MODFLOW (Harbaugh et al., 2000), and an optimisation model MODMAN (Greenwald, 1998) and LINDO (Schrage, 1990). MODMAN was used to generate the response matrix and to formulate the optimisation problem while LINDO was used to solve the linear optimisation problem. This is because the tools required to fully incorporate the nonlinear response of the unconfined aquifer into an optimisation problem were not developed or readily available. Most currently, Ahlfeld and Baro-Montes (2008) solved the unconfined groundwater management problem using a method of successive programming techniques. This solution method was incorporated in the United State Geological

Survey (USGS) groundwater management software package known as GWM (Ahlfeld et al., 2005) and available for use. Therefore, this software package was used to solve the optimisation problems which involve artificial recharge in the unconfined aquifer of the Samail Lower Catchment, Oman. In this study the use of multi-objective optimisation in artificial aquifer recharge study was also demonstrated, by applying Genetic Algorithms (GA) implemented in the NSGA-II by Deb et al. (2002). This approach allows multiple conflicting objectives to be addressed.

7.2.1   Study area

The study area is located on the southeast Batinah coast of Oman, about 50 km west of Muscat. It covers an area of about 60 km^2. The catchment is narrow in the upstream and spreads to form a delta fan closer to the coast. The mean annual rainfall varies between 300 mm in the highlands to about 90 mm on the alluvial plain (Kwarteng et al., 2009). The area has been an important source of water for potable water supply to a large part of the capital Muscat throughout the last three decades.

In order to harness surface runoff that occur in the wadis during heavy rainfall events, which subsequently can be used for effective recharge of the groundwater resources and prevent seawater intrusion, several recharge dams have been constructed in the different part of the country (Al-Battashi and Syed, 1998). The Al Khawd dam is one of the oldest and largest dams in Oman that was constructed in the Samail Lower Catchment around 7 km away from the Gulf of Oman (Figure 7.1). This dam was constructed between December 1983 and March 1985 to store wadi flow from the Samail Catchment and to enhance aquifer recharge by gradually releasing the stored water to the wadis downstream of the dam. However, given the aridity of the climate, the possible effect of climate change on rainfall amount and distribution, and the rapidly increasing water demand these recharge dams may not be enough to replenish or control over exploited aquifers. Therefore, interest is growing to use treated wastewater as an alternative source of aquifer recharge.

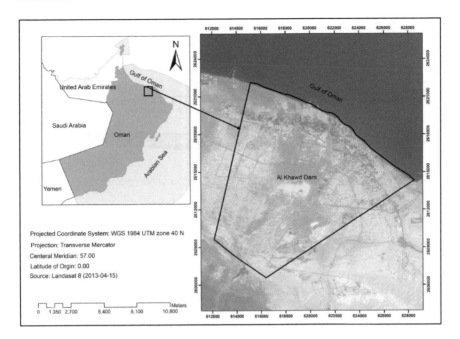

Figure 7.1: Satellite image of the study area

7.2.2 Geology of the study area

The aquifer is formed in predominantly alluvial sediments of different age and composition (mostly sand and gravel). The study area is mainly characterized by quaternary surface deposits. Tertiary sedimentary cover exists at the narrow upstream part of the study area. The ophiolites are exposed on both sides of the wadi Samail in the upstream boundary of the study area. According to Abdalla and Al-Rawahi (2012) and references cited therein, the lithology of the study area can be grouped in three different successive alluviums (Figure 7.2): (1) The upper gravel unit: predominately composed of large-size gravels including boulders, (2) The middle unit: discontinuous clayey gravel that is found in the form of lenses between the upper and lower gravel units, (3) The lower unit: cemented gravel that is more compacted and conglomeratic. The alluvium is over 300 m thick across most of the coastal plain (Pollock, 1994). The thickness of the aquifer gradually increases from the upstream to the coast of the sea. In this study, due to limited data availability, the thickness of the aquifer has been represented in a simplified manner. In the upstream end it is assumed to be 100 m and linearly increases towards the sea coast where it reaches a maximum value of about 300 m. The maximum drilled depth observed in geologic logs close to the upstream boundary support this assumption.

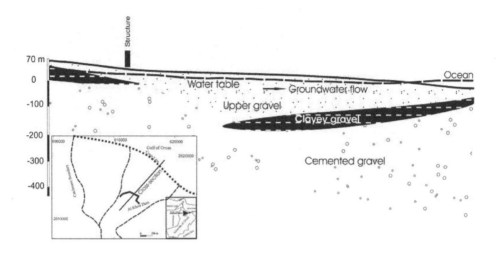

Figure 7.2: Geology of the study area (source: Abdalla and Al-Rawahi (2012))

7.2.3   Numerical model

Conceptual model

The model domain covers an area of 60 km². The conceptual setup of the model is based on limited data available and some previous studies. The aquifer has been conceptualized as a single hydrogeological layer. But, the single layer model was further divided into four homogenous computational layers for positioning of injection and recovery wells screens. The inflow to the model domain consists of recharge from precipitation and wadi flows, and subsurface flow from the neighbouring aquifer. The outflow includes discharge into the sea, existing pumping wells and evapotranspiration. The model was setup using MODFLOW code (Harbaugh et al., 2000) in the Processing MODFLOW, PMWIN modelling environment (Chiang and Kinzelbach, 1998).

Model Grid and Layering

The finite-difference mesh consists of 692 rows and 687 columns with a uniform grid size of 30 m x 30 m. The aquifer is considered as an unconfined aquifer with thickness increasing towards the sea coast. The elevation of the first layer was obtained from ASTER[1] 30 m x 30 m DEM (digital elevation model). The elevation of the top of the bedrock unit generally was unknown as no wells penetrate to bedrock. Therefore, the

---

[1] http://asterweb.jpl.nasa.gov/gdem.asp

aquifer is divided into four homogenous layers with thickness varying linearly from 100 m in the upstream to 300 m at the seacoast.

Boundary conditions

Three types of boundaries were used in the model (Figure 7.3): no flow (the east and west boundary of the model domain), constant head boundary (sea coast) and general head boundary (GHB) representing groundwater flow from the neighbouring aquifer from south. The two side boundaries are assumed to be no flow boundaries represented by streamlines. Although, flow cannot occur across streamlines, stresses can change the flow patterns and shift the position of streamlines. Therefore, the two lateral no flow boundaries were expanded, and the distance from the centre of the model to the no flow boundary has been chosen such that no influence would occur on the boundary due to the operation of the artificial recharge system.

Recharge boundaries

Recharge events within the wadis commonly occur as a result of channel flows associated with flood events. Recharge from precipitation and wadis were simulated as a specified flux boundary in the Recharge Package of MODFLOW and were applied to the topmost active layer. Seven recharge parameters and associated zones (Figure 7.3) were defined as follows: Rch1 corresponds to recharge from precipitation, Rch2 corresponds to recharge from wadi Samail upstream of the dam; Rch21 corresponds to recharge from wadi Samail upstream end, Rch3 corresponds to recharge from the recharge dam, Rch4 to Rch7 corresponds to recharge from wadis downstream of the dam from east to west respectively. Recharge estimation from the wadis is very challenging and often carries a lot of uncertainty. Initial value of recharge from precipitation was assumed to be 10% of the annual precipitation.

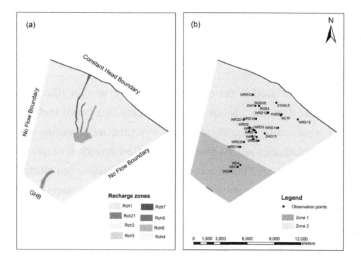

Figure 7.3: Recharge zones and boundary conditions (a), and hydraulic conductivity zones and observation point locations used in model calibration (b)

Groundwater withdrawal (existing pumping wells)

In this study 45 public water supply wells, two wells that supply water to oil company (PDO), and 67 irrigation wells were considered. The number of irrigation wells is based on an inventory study carried out in 1995. The well package of MODFLOW simulates a specified-flux boundary in each model cell to which a well is assigned based on the withdrawal rate for each well. Actual daily pumping rates were obtained by multiplying the yield of each well (available from existing records) with the average assumed pumping hours of a day (6 hours for public water supply wells, 9 hours for PDO wells and 4 hours for irrigation wells). With these pumping rates a total of 22,712 m^3/day withdrawal using public water supply wells was used as input to the model.

Evapotranspiration

Discharge by evapotranspiration from the model was simulated with a head-dependent function that decreased linearly with depth. To simulate discharge by evapotranspiration, maximum evaporation rate, elevation of the evapotranspiration surface and extinction depth need to be specified in the model. When the groundwater table is between the evapotranspiration surface and the extinction depth the extraction varies linearly with the groundwater table elevation. A maximum evapotranspiration rate is assumed to occur when the water table is at land surface. ETo calculator, a software developed by the Land and Water Division of FAO, Raes (2005) was used for calculating Reference evapotranspiration (ETo). The ETo calculator assesses ETo from meteorological data by means of the FAO Penman-Monteith equation. The maximum evapotranspiration rate computed by the program is about 5 mm/d. Based on literature study e.g. Shah et al. (2007) extinction depth was assumed to be 7 m below land surface.

Initial conditions and hydraulic properties

The observed steady state groundwater level distribution for the model was based on groundwater level measurements in the period 2007-2012. These groundwater levels are assumed to represent equilibrium conditions that existed after the withdrawal of large quantity of groundwater from the basin in the early 2000, but before the start of Barka desalination plant in 2003. Due to lack of subsurface information, the aquifer conditions were defined as homogenous. Two zones of different hydraulic conductivity (K) (Figure 7.3b) were used for zones of the aquifer dominated by cemented sand and gravel in the upstream part and loose gravel and sand formation in the other. In this model, the K values used were assumed to represent the regional scale effective properties of the bulk aquifer. A sensitivity analysis of this parameter is presented in the results and discussion section of this chapter.

Model calibration

Steady state flow model was calibrated using time averaged observed heads from 32 monitoring wells (Figure 7.3b) measured from January 2007 till October 2012. Hydraulic conductivity, GHB conductance, and recharge were used as model calibration parameters. During model calibration these parameters were adjusted until the simulated head matched the observed values. The calibration was performed using PEST. During the optimisation, observations were assumed to be of equal weight (equal importance in determining the optimisation outcome). Figure 7.4 shows the comparison of the measured and observed groundwater levels. In general, good correlations between the observed and simulated groundwater levels were obtained. The mean absolute error between the observed and simulated groundwater level values was 0.63 m. The calibrated model parameters are presented in Table 7.1. Table 7.2 presents the water budget of the calibrated model, while the groundwater head distribution is shown in Figure 4.

Table 7.1: Calibrated model parameters

Parameter	Values	Description
Hk1	0.79	Hydraulic conductivity of zone1 (m/d)
Hk2	7.76	Hydraulic conductivity of zone2 (m/d)
Rch1	3.64×10^{-6}	Recharge from precipitation (m/d)
Rch2	1.59×10^{-2}	Recharge from wadi upstream the dam (m/d)
Rch3	1.00×10^{-5}	Recharge from the dam itself (m/d)
Rch4	3.92×10^{-3}	Recharge from the downstream wadi (East) (m/d)
Rch5	5.28×10^{-3}	Recharge downstream wadi (m/d)
Rch6	8.25×10^{-5}	Recharge downstream wadi (m/d)
Rch7	9.88×10^{-5}	Recharge from the downstream wadi (West) (m/d)
Rch21	3.29×10^{-4}	Recharge from wadi far upstream the dam (m/d)
GHB	1.00	Conductance (m^2/d) for General Head Boundary
Vani1	10	Vertical anisotropy (m/d) for zone1
Vani2	1	Vertical anisotropy (m/d) for zone2

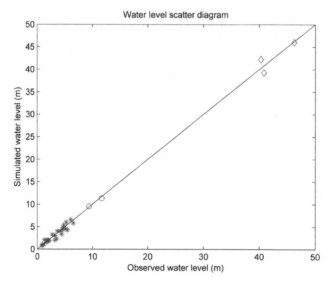

Figure 7.4: groundwater level scatter diagram (diamond points correspond to upstream points (WD1, WD4 and WD5), circle points correspond to the middle points (WRD19 and WRD20), and star points correspond to the downstream observation points)

Table 7.2: Water budget of the whole model domain

Flow term	In (m³/d)	Out (m³/d)
Constant head	412	12832
Wells	0	21343
Recharge	41944	0
ET	0	9067
GHB (groundwater inflow from south)	886	0
Sum	43242	43242

The most significant inflow to the model comes from recharge, with small contributions from the general head boundary (GHB) representing the inflow from the south and the constant head boundary representing the sea (Table 7.2). The inflow from the sea (indicating some seawater intrusion) is very limited and occurs in the north-western part of the model as indicated by the contours of groundwater head shown on Figure 7.5a. Most of the outflow from the model is via the existing pumping wells, followed by the components of evapotranspiration and outflow to the sea.

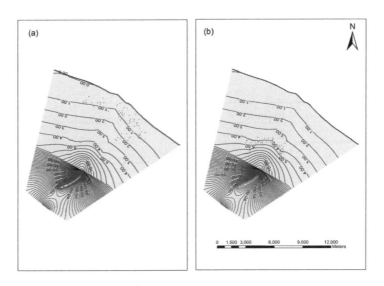

Figure 7.5: Steady state groundwater head distribution layer 1 (a) and layer 2 (b)

Although the calibration results seem to be relatively satisfactory, it needs to be emphasized that the developed model has several significant simplifying assumptions, most important of which are the introduction of only two homogeneous hydraulic conductivity zones, and a single inflow from the neighbouring aquifer (represented by the GHB to the south). The obtained pattern of groundwater heads presented in Figure 7.5 shows the influence of these two assumptions. Nevertheless, the obtained gradient of groundwater head (from upstream to downstream), as confirmed by the calibration results (Figure 7.4), gives some confidence that the developed model is overall representative of the groundwater flow in this aquifer. Further improvements of the model are certainly possible, but for this study that focuses on the methodology for simulation-optimisation of the artificial recharge scheme, the current model has been adopted for further analysis.

Transient Simulations

The same hydraulic properties of the aquifer determined from the steady state model were used in the transient simulation of annual injection and recovery scenarios. The data regarding hydraulic heads and historical pumping rates were not sufficient to allow calibration of transient model parameters. Transient model was built based on literature values of storage parameters (specific yield of 0.2 (Papadoulos, 1980), and specific storage of 10^{-4}). Water levels generated from the steady state model were used as initial water levels in the transient simulation. Table 7.3 shows the water budget of the whole model domain after 365 days of transient simulation. Figure 7.6 shows the head distribution in the aquifer after 365 days of transient simulation for layer 1 and layer 2.

Table 7.3: Water budget of the whole model domain at time step 8 (after 365 days simulations) without MAR

Flow term	In (m^3/d)	Out (m^3/d)
Storage	63	115
Constant head	412	12832
Wells	0	21343
Recharge	41944	0
ET	0	9084
Head dependent boundary	957	0
Sum	43375	43375

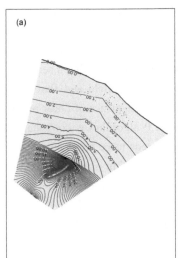
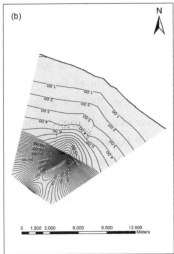

Figure 7.6: Head distribution at time step 8 stress period 2 (after 365 days simulation), layer 1 (a), and head distribution at time step 8 stress period 2 (after 365 days simulation), layer 2 (b)

7.2.4   Formulation of the Optimisation problem

The Ground-Water Management software package GWM (Ahlfeld et al., 2005) is used to solve groundwater management optimisation problems formulated as linear, nonlinear and mixed integer. In this study GWM was used to formulate the optimisation problem. The general formulation of a GWM optimisation process consists of three components: the objective function, decision variables and a set of constraints. In GWM three types of decision variables are supported: flow-rate decision variables (withdrawal or injection rate of a well), external decision variables (source or sinks of water that are external to the flow model), and binary variables (which have values of 0 or 1 and are used to define the status of the flow rate or external decision variables). A single objective function is supported by GWM which can be specified to either minimize or maximize the weighted sum of the three types of decision variables. The constraints supported in GWM are: (1) upper or lower bounds on the flow rate and external decision variables; (2) linear summation of the three types of decision variables; (3) hydraulic head based constraints, including drawdown, head differences, and head gradients; and (4) stream flow and stream flow depletion constrains.

GWM uses the well known Response Matrix Solution approach to calculate the change in head at each constraint location that results from the perturbation of a flow rate variable; these changes are used to calculate the response coefficients. The response matrix approach is applicable to linear system, notably confined aquifers. In unconfined aquifers, the changing saturated thickness produces nonlinearity in the groundwater system. The successive linear programming (SLP) solution algorithm implemented in GWM was used to solve the nonlinear optimisation problem by solving a series of sub problems, which take the form of linear programs (LP) (Ahlfeld and Baro-Montes, 2008; Ahlfeld and Mulligan, 2000). The idea is to solve a sequence of linear problems, iteratively till a convergence criterion is met. The SLP algorithm is computationally more intensive than LP because the Jacobian matrix needs to be recomputed at every iteration of the algorithm. More information about the algorithm is available in Ahlfeld and Baro-Montes (2008). Two groundwater management problems for transient conditions were formulated using GWM. To allow easy comparison between the optimisation algorithms of SLP and LP methods results obtained by both methods are presented in the results section.

Management problem 1: Four month of injection without recovery

The objective of this management problem is to maximize the total quantity of water injected into the aquifer for transient condition while satisfying head constraints that prevent excessive groundwater mounding. The objective function of this model can be written as Equation 7.1.

$$Maximize \ \sum_{n=1}^{N} QT_{Qn} \qquad\qquad (7.1)$$

Where:

Q is injection flow rate at managed well site, N is total number of injection wells, and T_{Qn} is the total duration of injection at well site.

The number and locations of potential injection well sites are fixed. These are introduced close to- but below the Al khawd dam. The wells are arranged in cluster mainly for operational reason. The decision variables are the rates of injection wells. The constraints are: (1) the maximum groundwater level is not allowed to go above 5 m below the ground surface; (2) the lower and upper bound of the injection constraints. Though from economic considerations, there exists a minimum injection rate below which the injection well is no longer cost-effective to operate, for this optimisation problem we used a lower bound of zero injection rate and upper bound of 1000 m^3/d.

Management problem 2: Four month of injection and eight month of recovery

The objective of the second management problem is to maximize the total quantity of water that can be recovered with an additional constraint related to the violation of the head gradient direction from the aquifer to the sea. A change in this gradient direction would result in seawater intrusion. The objective function of this groundwater management problem is similar to management problem 1 and is written as Equation 7.2.

$$Maximize \ \sum_{n=1}^{N} QwT_{Qn} \qquad\qquad (7.2)$$

Where:
Qw pumping flow rate at managed wells site, N is total number of extraction/recovery wells and T_{Qn} is the total duration of flow at well site.

The optimal injection rates determined from management problem 1 are introduced as input. The potential recovery pumping well sites (with fixed number and locations) are introduced down-gradient from the injection wells. The decision variables are the rate of these pumping wells. The constraints are (1) the head gradient at the control point along the sea coast should be greater than head gradient of the transient simulation pre-intervention; and 2) the lower and upper bound of the pumping rates; assumed to be zero and 1500 m³/d respectively.

In GWM model head gradient constraint can be imposed easily with the response matrix technique. However, with this model it is difficult to include directly the travel time between the injection and recovery wells as constraints. To include travel time constraint the simulation-optimisation model need to be incorporated with particle tracking methods. Therefore, the travel time constraint was not directly imposed in groundwater management problem 2. As was done in Jonoski et al. (1997), groundwater management problem 2 was solved in a two step procedure: First the optimisation problem was solved using gradient constraints only, and optimal recovery wells were obtained. Second, once the optimal recovery wells found, the particle tracking code, PMPATH (Chiang and Kinzelbach, 1998) was used to check the travel time constraint. By placing particle on each injection wells, close to the recovery wells forward particle tracking was carried out and the travel time of the fastest particle arrived to the recovery wells were compared with the regulatory limit or minimum travel time constraint between the injection and recovery wells which is 60 days.

Management problem 3: Four month of injection and eight month of recovery with travel time constraints

Management problem 3 is formulated as a multi-objective optimisation problem (also referred to as vector optimisation) characterized by simultaneous optimisation of multiple objectives. In this management problem two objectives were considered namely; Maximization of recovery rates and Maximization of the average groundwater level at specified control points along the sea coast. The decision variables and constraints imposed are also different from groundwater management problem 2. While in groundwater management problem 2 rates of recovery wells were considered as the decision variable, in groundwater management problem 3 both recovery rates and recovery well locations were used as decision variables. Also, in groundwater management problem 2 travel time constraint was not introduced directly in the formulated optimisation problem. But, in groundwater management problem 3, travel time constraint was directly included in the optimisation technique by coupling the simulation model including particle a tracking model, MODPATH (Pollock, 2012) to the NSGA II optimisation algorithm.

7.3   Results and discussions

In this study MODFLOW was used to simulate the flow field in response to injection and recovery of water. PMPATH and MODPATH were used to track the movement of imaginary particles by advection between the injection and recovery wells at the velocity calculated from heads generated by MODFLOW. GWM and Multi-objective Genetic algortim (NSGA II) were used to formulate and solve the optimisation problems based on MODFLOW simulations. The 365 days of simulation is assumed to be recycled every year. In the subsequent sections the optimisation results are presented and discussed.

7.3.1   Management problem 1

The effect of storage time on system performance is always of interest. For example, water applied adjacent to a point of natural discharge would increase discharge promptly, but might have little effect on storage. To maximize storage benefits, it is recommended that recharge sites need to be located far from the discharge points such as rivers, lakes, springs. Similariy, when hydraulic gradient around the injection well is sufficiently steep, and the aquifer hydraulic conductivity is sufficiently high there is real concern that the stored water will rapidly move away from the well prior to or during recovery. Therefore, in this study we selected injection site based on aquifer porperty, its distance from the sea, and distance to exising public water supply wells. The decision variables for this management problem are the injection rates at 109 potential injection wells shown in Figure 7.7a.

In addtion to the location of the wells, the spacing between them is an important factor. Pyne (1995) recommended to use 100 to 300 m spacing between the ASR wells. If the well spacings are lesss than this diameter, he argued that the storage cones from adjacent ASR wells tend to combine. Based on this recommendation we adopted a spacing of 210 m between injection wells. The maximum groundwater mound in the aquifer usually occurs at the potential injection well sites. Therefore, head constraints at each potential injection well site were used during the optimisation. The groundwater levels at these locations were constrained so that none will exceed their allowable limits specified as 5 m below the ground surface.

Using the SLP optimisation algorithm, out of the potential 109 well sites (Figure 7.7a), only 68 injection wells (Figure 7.7b) with non-zero injection rates were selected in the final solution. The total optimal injection rate was computed to be 66595 m^3/day. The total volume of injected water during the four month period is about 8 x 10^6 m^3. The total injection rate computed using LP was 66382 m^3 per day, which is slightly less than injection rate obtained using SLP. The water budget for this scenario at the end of injection period is presented in Table 7.4 (SLP results). For the same optimal solution Figure 7.8 shows the head distribution in the aquifer after one month and four month of injection.

Table 7.4: Water budget of the whole model domain time step 4 stress period1 (end of injection period) SLP

Flow term	In (m^3/d)	Out (m^3/d)
Storage	125	59388
Constant head	408	19586
Wells	66595	21343
Recharge	41944	0
ET	0	9641
General Head Boundary	886	0
Sum	109958	109958

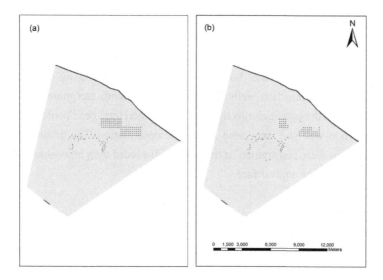

Figure 7.7: 109 potential injection wells arranged in a grid at spacing of 210 m (blue), located in the second layer (red are existing public water supply wells) (a), and location of 68 optimal (SLP) injection wells (blue) and existing public drinking water supply wells (red) (b)

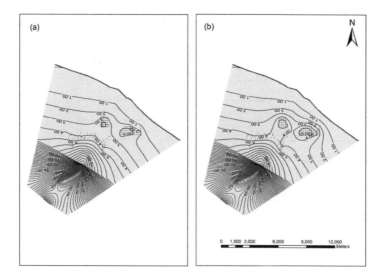

Figure 7.8: Head distribution obtained using optimal well of scenario 1 after 30 days of injection (SLP), 2nd layer (a), head distribution obtained using optimal well of scenario 1 after 120 days of injection (SLP), 2nd layer (b)

7.3.2   Management problem 2

The objective of this management problem is to maximize the total quantity of water that can be recovered without violating head gradient constraints at the sea coast. Gradient constraints were placed at every six pair of grid nodes along the coast. In total 58 hydraulic gradient locations were used during the optimisation. Since the interest is to recover the same water that was used for recharge, the recovery wells were located down-gradient from the injection wells. If recovering the same water is not important, it is possible to increase the recovery efficiency by placing additional recovery wells upstream of the injection facility. In this groundwater management problem optimal injection rates and well locations from management problem 1 were used and optimal rates and locations of recovery wells were optimized. The number of initial potential location well sites for recovery was 54, all located down-gradient of the injection site, as shown in Figure 7.9a. These wells are placed in the first layer of the model domain. Out of the 54 potential recovery wells, 16 wells with a total recovery rate of 21448 m^3/day were found to be optimal (Figure 7.9b). The total recovered volume of water in the eight month period is about 5.3 x 10^6 m^3. These values were obtained using the SLP algorithm. Applying the LP solution led to a slightly higher total recovery rate, which is 21531 m^3/day. Table 7.5 shows the water budget of the whole model domain after 365 days of transient simulation for the four month of injection and eight month of recovery scenario. Figure 7.10 shows the head distribution in the aquifer at the end of the eight month recovery period for layer 1 and layer 2.

Since, it is not possible to directly impose travel time constraint in the optimisation problem formulation using GWM the travel time constraint was checked using PMAPTH, a particle tracking model. By placing particles on injection wells, which are close to the optimal recovery wells forward particle tracking was carried out and the travel time of the fastest particle arrived to the recovery wells or existing irrigation wells were compared with the regulatory limit, which is 60 days. From this analysis it was observed that the fastest particle travel from the injection well to the existing irrigation well in about 61 days. None of the optimal recovery wells capture injected water less than 60 days old.

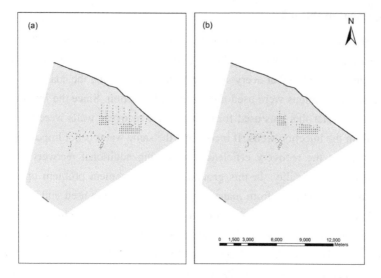

Figure 7.9: Optimal injection wells and 54 potential recovery well locations (a), and optimal injection and 16 recovery well locations (b)

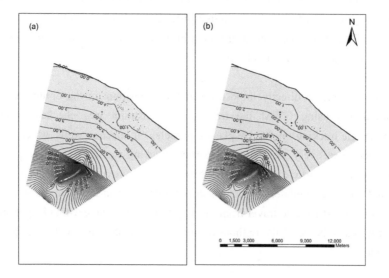

Figure 7.10: Head distribution at time step 8 stress period 2 (at the end of the eight month recovery period), layer 1 (a), and head distribution at time step 8 stress period 2 (at the end of the eight month recovery period), layer 2 (b)

7.3.3   Management problem 3

In groundwater management problem 2 only the rate of recovery wells were used as decision variables. However, it could be possible that a different arrangement of wells could satisfy all the constraints with potentially higher recovery rates. This possibility was examined further by repeating the optimisation problem 2 using recovery rate and well locations as decision variables. This was done using multi-objective optimisation. The multi-objective optimisation problem was solved using NSGA-II.

We applied NSGAII implemented in MATLAB to solve management problem 3 with two objectives: maximizing the total recovery rates and maximizing the average water level at grid cells located next to the sea. The two objectives conflict with each other. Since Genetic algorithms do not have explicit constraints; constraints are handled via the fitness function-penalty for violation. An improvement in any one objective can only be achieved at the expense of sacrificing the objective value of the other one; meaning there is always a trade-off. From the Pareto-optimal points; one may prefer one point over the other depending on the situation. In our multi-objective genetic algorithm, the number of population size was specified as 100, and crossover and mutation probabilities were specified as 0.9 and 0.1, respectively. Figure 7.11 shows all nondominated solutions after 20 generations.

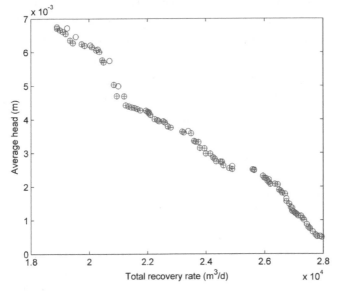

Figure 7.11: Obtained nondominated solutions (circles with red plus sign denotes rank1 Pareto-optimal points)

Controlling further seawater intrusion during recovery

In groundwater management problem 2 head gradient constraints along the sea coast were used to keep positive gradient towards the sea, similarly in groundwater management problem 3 heads at specified location along the sea coast were maximized by considering their average value as one objective function. However, these can protect only wells located sufficiently far from the sea coast. Wells located near the sea coast may be close to the freshwater-saline water interface and become subject to upconing (withdrawal of brackish or saline water) even when the horizontal gradient of groundwater heads is kept in the direction towards the sea. These effects were not included in the formulation of the optimisation problems. However, after the optimal solutions were obtained the absence of upconing effects was verified by running the simulation model using the optimal solution and then checking the water budgets of the whole model domain.

Comparing the water budget of the transient simulation before any intervention (Table 3) and transient simulation after intervention, at the end of the four month injection and eight month of recovery scenario of groundwater management problem 2 (Table 7.5), we can observe that the discharge flowing into the model domain from the constant head boundary is decreasing. The discharge from the constant head (from the sea) is decreased by about 19 m^3/d, while the outflow to the sea is increased by about 856 m^3/d. In addition, there is a significant increase in storage in and out of the model domain. Similar analysis was also carried out for groundwater management problem 3 (not shown). Based on these observations it can be concluded that the extraction wells are pumping at reasonable rate and are not causing further seawater intrusion by upconing. However, density dependent flow modelling needs to be carried out in further studies in order to understand better the seawater-freshwater interactions under MAR.

Table 7.5: Water budget of the whole model domain time step 8 stress period 2 (at end of the eight month recovery period) SLP

Flow term	In (m^3/d)	Out (m^3/d)
Storage	23774	1410
Constant head	393	13688
Wells	0.0	42791
Recharge	41944	0.0
ET	0.0	9108
General Head Boundary	886	0.0
Sum	66997	66997

7.3.4   Recovery efficiency

Recovery efficiency is defined as the percentage of the water volume stored that is subsequently recovered while meeting a target water quality criterion in the recovered water (Maliva and Missimer, 2010; Pyne, 1995). Recovery efficiency usually has little significance where both stored water and native water are potable. However, when there is significant water quality difference between stored and native water mixing has to be controlled and recovery efficiency becomes an increasingly important factor in the assessment of the ASR feasibility. The total recoverable volume is a function of the injection and recovery rate of the system, the duration of injection, and system recovery efficiency. Total injection in management problem 1 is 8 x10^6 m^3 and the total recovery volume considering management problem 2 is 5.3 x10^6 m^3. This means the recovery efficiency is about 66%. For groundwater management 3, the recovery efficiency depends on the choice of recovery rate from the nondominated solutions (Figure 10). What is important is that the water that is not recovered from storage in this aquifer ultimately benefit the study area, since the lost water will recharge the aquifer, prevent salt water intrusion and may eventually be further considered for abstraction through private wells by the farmers.

7.3.5   Sensitivity Analysis

Sensitivity analysis was performed to examine the overall responses of simulation-optimisation results to changes in hydraulic conductivity. This information is useful to gain understanding of how uncertainties in hydraulic conductivity affect the model results. Table 7.6 presents the sensitivity of injection and recovery scenarios. The 3rd and 4th columns respectively in Table 7.6 show the optimal number of injection and total quantity of water to be injected for different values of hydraulic conductivity for groundwater management problem 1. Similarly, the 5th and 6th columns respectively shows the optimal number of recovery wells and total quantity of water to be recovered for different values of hydraulic conductivity for groundwater management problem 2. All results are obtained using the SLP optimisation. The sensitivity analysis results show that though an increase in hydraulic conductivity helps to inject more water to the aquifer, the recovery rates decrease with increase in hydraulic conductivity.

Table 7.6: Sensitivity of injection and recovery scenarios to changes in hydraulic conductivity

Hydraulic conductivity (m/d) for zone 1	Hydraulic conductivity (m/d) for zone 2	Optimal number of injection wells	Total quantity of water injected (m^3)	Optimal number of recovery wells	Total quantity of water recovered (m^3)	Recovery (%)
0.4	4	67	7.8×10^6	22	7.0×10^6	90
1.6	16	78	9.3×10^6	13	3.6×10^6	39
2.4	24	85	1.0×10^7	9	3.0×10^6	29
3.2	32	92	1.1×10^7	10	2.6×10^6	24

7.4 Conclusions

A simulation-optimisation modelling approach was used to investigate the feasibility of MAR in the Samail Lower Catchment, Oman. The modelling framework involved the coupling of simulation-optimisation approach, and particle tracking algorithm (PMPATH/ MODPATH). The work presented herein demonstrates the use of groundwater simulation and optimisation to construct a decision support system for solving a groundwater management problem associated with MAR in the Samail Lower Catchment. The simulation-optimisation results successfully addressed groundwater management problems that involve maximization of groundwater recharge and maximization of its extraction/ recovery. System, operational and regulatory constraints were included in the modelling framework. The simulation-optimisation study presented in this study demonstrated the feasibility of MAR. It is shown that the total injection rates could be as high as 8×10^6 m^3 for management problem 1, and the total recovery volume considering management problem 2 is 5.3×10^6 m^3. The recovery efficiency for the whole period is about 66%. The multi-objective algorithm is able to derive the entire nondominated front of the solutions for groundwater management problem 3. Overall, the simulation-optimisation approach forms the basis for making decisions regarding the location, and injection and recovery rates of each optimal well. However, it should be noted that due to known non-uniqueness problem the calibrated parameters are by no means unique; numerous combination of parameter can give the same result and consequently affects the simulation-optimisation results.

MAR using treated wastewater would increase the availability of water that could be used for agriculture and domestic uses and decreases the need to develop more capacity of desalination systems. The system was designed so that there is sufficient travel time between recharge and recovery wells so that to allow inactivation of pathogens or other nutrients commonly present in reclaimed water. Water particles residence times in the aquifer are more than the regulatory requirement before they are recovered/ extracted. However, further water quality issues of recharged water need to be studied before final decisions are made.

8 Conclusions and Recommendations

8.1 Conclusions

Groundwater is one of the major sources of drinking water supply for many countries in the world. It also constitutes much of the world's irrigation water supply, and is important for many ecosystems as these ecosystems are closely related to shallow groundwater. There is growing body of evidence indicating that groundwater resources in many places are under increasing threat due to contaminations and over-use. Increase in demand in future due to population increase and climate change may also challenge groundwater management. To ensure the functioning of the groundwater systems in the future, sustainable management considering its quantity, quality and its relationship with surface water systems need to be developed. Simulation models provide a useful way to investigate the response of the aquifer to different management options and to understand GW-SW dynamics. However, simulation models alone may not enable the groundwater manger to define the optimal planning or management options satisfying all the hydrologic and environmental constraints. On the other hand, coupled simulation-optimisation approach enable groundwater manger to find optimal solutions of groundwater in an efficient manner.

This thesis illustrates the application of different simulation and simulation-optimisation modelling approaches that can help to obtain a better understanding of the complex contaminant transport mechanisms, groundwater-surface water interactions and ways to manage complex groundwater systems. In the subsequent sections a summary of important conclusions drawn from the different application case studies is given.

Assessing the feasibility of natural attenuation as a remedy for chlorinated aliphatic hydrocarbons contaminated groundwater.
To predict the future migration of chlorinated aliphatic hydrocarbons (CAHs) dissolved in groundwater at the Vilvoorde-Machelen site, a numerical groundwater flow and contaminant transport model was developed for the site using MODFLOW and RT3D. After calibration and sensitivity analysis, the model was used to evaluate the fate and transport of dissolved CAHs in site groundwater under the influence of advection, dispersion, adsorption, and biodegradation. Model predictions were made for two scenarios; (i) a base line scenario considering natural attenuation, and (ii) a scenario that involves source zone removal and natural attenuation. The purpose of the forward modelling is to predict how CAHs plumes move away from the source as a function of

141

time. This approach is used to explain how a plume of contamination evolved and how it will behave in future. It also helps to evaluate the migration of the plume towards the identified receptors and to design and compare different remedial alternatives.

Natural attenuation scenarios in this study were evaluated using the Lines of Evidence approach proposed by Wiedemeier et al. (1998), which includes reduction of contaminants mass at the site and presence and distribution of geochemical and biochemical indictors of natural attenuation. Results of the study using natural attenuation scenarios revealed that chlorinated solvent concentrations are reducing along the flow path. Degradation products such as cis-1-2-DCE, VC, are observed in the groundwater (where the ratio of the daughter products to parent solvents is increasing over time and cis-1,2-DCE is the predominant DCE isomer). All these factors are related to specific natural attenuation processes and hence are good indicators that natural attenuation is taking place. Natural attenuation is occurring at the site and the estimated biodegradation rates are large enough to reduce the concentration of PCE, TCE and DCE to remediation standard. However the degradation rate under natural attenuation may not be enough to lower the VC concentration to remediation standard. Based on these observations natural attenuation is not recommended as the only remedial measure for removing VC at the site.

Results of the study using combined natural attenuation and source zone removal show about 21 % more reduction in VC concentration than natural attenuation alone. The assumption of source zone removal is that source zone contaminant concentration is successfully reduced by removing the source and once the source is removed sufficient removal can be achieved by the natural attenuation processes. This scenario significantly reduced the concentration of the contaminant near the source but less close to the river.

In conclusion, natural attenuation as well as natural attenuation with source removal are sufficient to reduce the concentration of PCE, TCE and DCE to remediation standard. But they are not enough to lower the VC concentration at the site to remediation standard. Hence, other alternatives such as enhanced biodegradation should be investigated further to reduce VC.

Local scale groundwater-surface water interactions
There has been an increased interest in further characterizing local scale GW-SW interaction in the Zenne River, Belgium. The reason for this increased interest is a rising appreciation that the hyphoreic zone played an important role in CAHs attenuation.

Often groundwater flow models such as MODFLOW are used to simulate GW-SW interactions. These models work best when they are applied at the regional scale; however their relative utility can be questionable if they are used to address GW-SW interactions at such smaller spatial scales.

In this thesis we used temperature as a natural tracer to quantify GW-SW fluxes exchange. The fluxes exchange between the Zenne River and an adjacent aquifer was modelled using the variably saturated two dimensional groundwater flow and heat transport model (VS2DH) from the USGS. In order to investigate the effect of temporal resolution of water level and temperature data on model performance and fluxes exchange dynamics, first the model was calibrated and validated using hourly time scale water level and temperature measured data. Secondly, the hourly data were aggregated into daily time scale and used for model calibration and validation. In both cases the model is able to reproduce temperature time series observed at multiple depths below the river bed. However, only the hourly model was able to capture the dynamics of river-aquifer fluxes during high river flow events. During high river flow events the river changed from gaining to losing river. The daily model is unable to capture these phenomena.

Therefore, when using models to simulate GW-SW interaction at a local scale it is crucial that special attention be given to the temporal resolution of the collected data for model calibration and validation.

Groundwater-Surface water interactions in a drained riparian wetland

This study investigated GW-SW interactions in riparian wetland Kielstau, North Germany in a setting of riparian wetland, drains and a river. The GW-SW interaction between a riparian wetland and river was quantified through numerous modelling efforts. Both numerical and analytical models were developed to calculate the flux rates between these two systems based on water level measurements and were then compared. The objective of this part of the research was to develop a conceptual model of the flux dynamics so that residence times for this site can be better quantified and then used for future nutrient studies.

The transient numerical groundwater model (MODFLOW) was setup for the riparian wetland using weekly water level data observed in spatially distributed wells in the riparian wetland. The MODFLOW model was used to quantify the residence time of the shallow groundwater flow, riparian wetland-river fluxes exchange and the impact of the existing drainage ditches on the overall water budget of the riparian wetland. The fluxes between the wetland-river systems were also quantified at a daily time scale using the analytical profile model (STWT1) for one cross-section of the riparian wetland-river system. Then, the flux exchange estimates obtained using the transient numerical groundwater model at the location where the analytical model was implemented was compared to the flux exchange estimates using analytical profile model.

Results obtained from the MODFLOW model showed that: (i) the river always act as a gaining river (ii) total outflow into the drainage ditches is about 64% of the total recharge into the riparian wetland, (ii) similarly, total outflow through riverbed accounts about 12% of the total recharge into the riparian wetland, and the existing drainage ditches considerably modified the groundwater flow paths. Results of the study obtained using the STWT1 model showed that the river is changing from a gaining river to a losing river many times during the simulation, particularly during high flood events. This provides an insight for the rate of flow from the river to the aquifer, which was not obtained with the numerical model due to limited availability of data at higher resolution.

It can be concluded that detailed knowledge of the velocity field and residence time would have significant implications for further studies related to nutrient transport in the riparian wetland. However, as observed in the analytical model the transient numerical groundwater model provides little information about the GW-SW interaction at the river-wetland interface. Near the river, the groundwater flow is strongly influenced by the river stage. The rise and fall of the river creates a dynamic zone between the two flow systems, thus influencing the flow patterns. Hence, transient models based on daily to sub-daily measurement data may be required to understand nutrient dynamics at this interface.

Optimal management of shallow groundwater levels in drained riparian wetlands
This study investigated the use of a combined simulation-optimisation approach for estimating optimal water level in drainage ditches used to remove excess water from the riparian wetland. The studied riparian wetland, Kielstau is located in the lowland areas of North Germany. In order to remove the excess water from the riparian wetland, open drainage ditches were used for decades. Currently, there are 11 drainage ditches that drain this small riparian wetland. The drainage ditches are located perpendicular to the Kielstau River, and are of different age, type and length. The shallow groundwater flow through the riparian wetland soils needed to have sufficient contact and residence time within the soil matrix for nutrient removal. The objective of this study was to find optimal water levels in the existing drainage ditches so that shallow groundwater flow

144

into the drainage ditches is minimal while providing necessary trafficability (e.g., grass cutting machineries) during wet periods. The rationale behind this is to allow minimum drainage outflow from the riparian wetland in order to allow sufficient residence time to enhance the nitrate removal capacity of the wetland by keeping the shallow groundwater level in the riparian wetland as high as possible.

Two global optimisation algorithms known as genetic algorithm (GA) and controlled random search (CRS4) together with the simulation model (MODFLOW) were used to solve the simulation-optimisation problem. The objective function was formulated as minimizing the total discharge outflow considering water level in the drains as decision variables, imposing the minimum water levels to allow trafficability during wet periods as constraints. For this site trafficability during wet periods was assumed to be achieved by lowering the water level at least 25 cm below the ground surface. The optimal drain water level elevations obtained by using the two global optimisation algorithms were found to be very close, which may indicate that the obtained drain water levels are near-optimal solutions. The results showed that in order to allow trafficability during wet periods the water levels in the drains need to be lowered from the current conditions. The simulation-optimisation approach provides an improved understanding as to how to manage the drainage water level for controlling shallow groundwater levels in the drained riparian wetland, particularly during wet periods.

Assessing the feasibility of managed aquifer recharge
In many coastal aquifers excessive pumping results in seawater intrusion problem. Due to excessive pumping, saline waters from the sea are migrating into the freshwater aquifer diminishing the availability and quality of the groundwater source. Therefore, restorations of groundwater in these areas have become an important component of groundwater management. Managed aquifer recharge (MAR) may be used as groundwater management option to replenish the aquifer. Managed aquifer recharge is practiced widely to store water during periods of surplus and withdraw during deficit from the aquifer and to mitigate seawater intrusions problem in coastal aquifers.

To assess the feasibility of MAR in the Samail Lower Catchment, Oman a simulation-optimisation modelling approach was used. Three groundwater mangment problems were considered: (1) four months of injection without recovery, (2) four months of injection and eight months of recovery, and (3) four months of injection and eight months of recovery with travel time constraint. The first two management problems were solved by GWM that couples the simulation model MODFLOW with successive linear optimisation solver. The objective function used were maximizing the amount of injected water for management problem 1 and maximizing the recovered amount of water in the second management problem. Head and gradient constrains were used in the optimisations. The third management problem was solved by coupling NSGA-II, MODFLOW and MODPATH. Coupling of MODPATH allows travel time constraint to be included directly in the optimisation. Two conflicting objectives; were maximizing the total recovery rate and maximizing the average water level near the sea coast were considered.

The objective of this study was to find the optimal location of injection and recovery wells with pumping rates that would maximize the injection and recovery rates. After the optimal injection and recovery wells location and rates were identified using GWM, particle tracking was completed using PMPATH in order to check that the travel time between the injection and recovery wells is not less than 60 days, as required for inactivation of pathogens or other nutrients commonly present in treated wastewater. Results of the study showed that the total injection considering groundwater management case1 could be as high as 8×10^6 m^3 in the four month period. Whereas the recovery amount for groundwater management in case 2 could be 5.3×10^6 m^3 during the eight month period. The total recovery efficiency then is about 66%. Result of the study using the multi-objective algorithm is able to derive the entire nondominated front of the solutions for groundwater management problem 3. Recovery efficiency for this management problem depends on the choice of recovery rate from the nondominated solution.

In conclusion, MAR using treated wastewater would have a real benefit for the region. It would increase the availability of water that could be used for agriculture and domestic uses. It also decreases the need to develop more capacity of desalination systems. The aquifer is capable of absorbing large volumes of water and the recovery efficiency is high enough to consider MAR as a feasible option. In addition, the unrecovered volume of water has benefit in controlling the landward migration of seawater. But, for further investigations density-induced flow should be considered among other water quality aspects.

Overall, the thesis has demonstrated the possibilities and limitations of combining different modelling techniques for several specific groundwater management problems Table 8.1 present the advantages and limitations of different modelling approaches which offer further guidance to water resource managers engaged in groundwater quality and quantity management.

Table 8.1: Problems analysed, characteristics of the problem and modelling approaches used

---

Natural attenuation as remedial option for chlorinated solvent contaminated groundwater

Characteristics of the problem
- Multi-species reactive transport of CAHs
- Multi-layer aquifer system
- Limited data is available

Modelling approach used and recommended
- MODFLOW and RT3D

Advantages
- Able to track the degradation of parent compound through its daughter products.
- Better understanding of CAHs fate and transport behaviour in space and time.
- Better prediction scenario for assessing remedial actions for CAHs.

Disadvantages
- More data is need for defining and calibrating the model
- RT3D model considers only single-phase flow, so the model may not be appropriate for simulating transport of contaminants that exists in a free phase separated from dissolved phase.
- Detailed site characterization is required.
- High computational cost.

Quantifying local scale GW-SW interactions

Characteristics of the problem
- GW-SW interactions are spatially and temporally varying and difficult to estimates
- GW-SW interaction on a meter or sub-meter scale on a given stream reach is required

Modelling approach used and recommended
- VS2DH, temperature as a natural tracer

Advantages
- Relatively less data requirement, enable the use of improved sensor technologies such as the use of fibre optic distributed temperature sensor data at meter scale spatial and measure cycle of minutes or hours.
- Gain new basic insights into stream-aquifer interaction
- Less expensive method to quantify stream-aquifer interaction
- VS2DH model can analyze problems in both one and two dimensions (longitudinal or cross-sectional)

Disadvantages
- Head data which is used in conjunction of temperature measurement often may not available or cannot be measured easily as temperature.
- Joint inversion of head and temperature data are needed to obtain better estimates.

Quantifying GW-SW interaction in a drained riparian wetland

Characteristics of the problem
- Groundwater flow velocities, residence time and flow paths, stream-aquifer exchange need to be determined.
- Limited water level data at different temporal resolution is available

Modelling approach used and recommended
- Combined modelling approach which involve the analysis and utilization of different data sets (numerical model (e.g., MODFLOW) and Analytical profile model at river cross-section (e.g., STWT1))

Advantage
- Provides a convenient way to utilize available data

Disadvantage
- Comparison of results may not be straight forward

Optimal management of shallow groundwater level in a drained riparian wetland

Characteristics of the problem
- Drainage systems increase loss of nitrate to receiving water bodies
- Drainage ditches that allow minimum drain outflow and trafficablity is desired

Modelling approach used and recommended
- Coupled simulation-optimisation approach
- Global optimisation techniques (e.g., GA, CRS4) coupled with simulation model MODFLOW

Advantages
- Allows non-linear problem of the unconfined aquifer to be solved efficiently.
- The approach has better performance than the design alternative using simulation model alone.
- The methodology can be applied in areas where agricultural drains are provided to remove excess water from water logged soils where high water levels restricts plant growth, and thus crop production
- Also allow proper spacing of drains in drainage systems in water logged agricultural soils.

Disadvantage
- Computational cost is high

Assess the Feasibility of Managed Aquifer Recharge

Characteristics of the problem
- MAR schemes design for a given management objectives and set of constraints
- The problem contains non-linear response because of the unconfined condition, however, the saturated thickness of the unconfined aquifer is large and the response is mildly non-linear.

Modelling approach used and recommended
- GWM (successive linear programming algorithm (SLP) coupled with MODFLOW)
- NSGA-II coupled with MODFLOW (multi-objective optimisation)
- PMPATH/MODPATH (particle tracking algorithms)

Advantage
- GWM is less computation time compared to global optimisation techniques.
- NSGA-II allows multi-objective problems with conflicting objective to be solved.
- NSGA-II also allows including travel time constraint in the optimisation process and it allows using the well locations as decision variables.
- NSGA-II is a global multi-objective optimisation techniques meaning global optimum can be found than local solutions.

Disadvantage
- In GWM, an option to include travel time as a constraint, which is an important parameter for the aquifer storage and recovery, is not available.
- The main disadvantage of NSGA-II its high computational cost, but this can be resolved by using parallel computing.

8.2   Recommendations

This thesis has demonstrated the use of groundwater models through different application case studies. Simulation and simulation-optimisation models were used to understand and mange the impacts of human activities on groundwater resources. These include understanding the flow of groundwater and contaminants at chlorinated solvent contaminated sites, and assessing remedial options, understanding GW-SW interaction at different scales, shallow groundwater level management through optimal drain water level control, and feasibility of managed aquifer recharge to replenish depleted aquifer. The real benefit of groundwater models is that they provide development scenarios in a quantified manner which can help decision makers in examining different groundwater management options. However, these models are only as good as the input data used in those models. When appropriate field data are lacking the application of these models is subject to considerable error. The approach used in this thesis presented several options for further improvement and research. The following recommendations give some aspect for further research.

Uncertainty analysis: There are many sources of uncertainty that impair contaminant transport model application. These include: measurement errors associated with imprecise collection, analysis and interpretation of data samples; sampling error due to having limited spatial and temporal data on which inference is made; and model structure error due to inaccuracies in the underlying conceptual models and how they represent physical, chemical, and biological processes. In general, it is very difficult to get sufficient and reliable data for contaminant transport modelling. Clement (2011) reported that one of the major problems that limits the use of reactive transport models at chlorinated solvent sites is the lack of problem-specific information of input parameters. There is a relatively high level of uncertainty associated with input data. Often there is lack of information about the location of sources, how much contaminant was released and when, and the present distribution of the contaminant. Though these past information are critical for reconstructing historical plume concentrations they are usually difficult to obtain. Recognizing all the limitations in contaminant transport modelling Konikow (2011) in his review paper "The secret to successful solute-transport modelling", concluded that the secret is simply to lower expectations.

Uncertainties also exist in hydrogeological systems. These uncertainties generally emerge from heterogeneities in the hydrogeological environment and scarcity of related data, and they can be further related to aquifer characteristics and/or physical, chemical and biological properties of the contaminant being released and transported (Qin and

Huang, 2009). Another source of uncertainty is uncertainty due to model conceptualization of the aquifer system.

Even if understanding and quantifying all sources of uncertainty is critical to the success of simulation modelling, in this thesis uncertainty analysis was not carried out at great level of detail, due to the heavy computational time involved with three dimensional contaminant transport modelling. However, the use of new technologies such as cluster computing, cloud computing, and parallel computing may ease the computational burden.

Other recommendations

- It is important that further contaminant source characterization is carried out on the Vilvoorde-Machelen site before remedial measures can be suggested;
- Spatially and temporally distributed measurements are required to reliably model plume evolution in space and time;
- The local scale GW-SW interaction in the Zenne River was carried out considering half of the river section. It is important that the full river cross section is considered. For this, it is essential to install monitoring wells on the left side of the river as well, along the existing cross section orientated perpendicular to the river;
- GW-SW interaction in the drained riparian wetland of Kielstau, North Germany was carried out only during winter conditions due to data limitation. But, it is important to consider seasonal fluctuations in the groundwater flow patterns as well;
- For the MAR application case study in Oman, it is important to estimate recharge using a water balance model or other available techniques. Other methods such as electrical resistivity survey in conjunction with bore logs should be used to reliably characterize aquifer geology and define the vertical distribution of the model layers;
- This study considers only hydraulic issues to assess the feasibility of managed aquifer recharge, but future studies should also consider water quality aspects;
- Injection wells may lead to rapid clogging - the likelihood of clogging is reduced when the injected water is of high quality;
- Field scale experiments, such as injecting water in the aquifer and monitoring the response of the aquifer from observation wells, help to better estimate aquifer parameters.

References

Abarca, E., Vázquez-Suñé, E., Carrera, J., Capino, B., Gámez, D., and Batlle, F., 2006, Optimal design of measures to correct seawater intrusion: Water resources research, v. 42, p. W09415.

Abdalla, O. A. E., and Al-Rawahi, A. S., 2012, Groundwater recharge dams in arid areas as tools for aquifer replenishment and mitigating seawater intrusion: example of AlKhod, Oman: Environmental Earth Sciences, p. 1-12.

Abriola, L. M., 1987, Modeling contaminant transport in the subsurface: An interdisciplinary challenge: Reviews of Geophysics, v. 25, no. 2, p. 125-134.

Ahlfeld, D., 1990, Two-stage ground-water remediation design: Journal of Water Resources Planning and Management, v. 116, no. 4, p. 517-529.

Ahlfeld, D. P., Barlow, P. M., and Mulligan, A. E., 2005, GWM--A ground-water management process for the US Geological Survey modular ground-water model (MODFLOW-2000), US Department of the Interior, US Geological Survey.

Ahlfeld, D. P., and Baro-Montes, G., 2008, Solving unconfined groundwater flow management problems with successive linear programming: Journal of Water Resources Planning and Management, v. 134, no. 5, p. 404-412.

Ahlfeld, D. P., and Mulligan, A. E., 2000, Optimal Management of Flow in Groundwater Systems, Academic press: AHarcourt Science and Technology Company.

Ahmed, M., Al-Maktoomi, A., Ebrahim, G. Y., Slim, Z., and Al-Jabri, S., 2013, Storing water underground: A response to anticipated water shortage due to climate change, The threast to Land and Water Resources in the 21st century: Mitgation and Resouration , September 4 -7: Le Meridien Chinag Rai Resort, china.

Al-Battashi, N., and Syed, R., Artificial recharge schemes in water resources development in Oman, in Proceedings Proceedings of the 3rd international symposium on artificial recharge of groundwater-TISAR1998, Volume 98, p. 231-236.

Al-Rashed, M. F., and Sherif, M. M., 2000, Water resources in the GCC countries: an overview: Water Resources Management, v. 14, no. 1, p. 59-75.

Alexander, M. D., and Caissie, D., 2003, Variability and Comparison of Hyporheic Water Temperatures and Seepage Fluxes in a Small Atlantic Salmon Stream1: Ground Water, v. 41, no. 1, p. 72-82.

Allan, J. D., 1995, Stream ecology: structure and function of running waters, Kluwer Academic Pub.

Anderson, M. P., 2005, Heat as a ground water tracer: Ground Water, v. 43, no. 6, p. 951-968.

Anibas, C., Fleckenstein, J. H., Volze, N., Buis, K., Verhoeven, R., Meire, P., and Batelaan, O., 2009, Transient or steady state? Using vertical temperature profiles to quantify groundwater–surface water exchange: Hydrological Processes, v. 23, no. 15, p. 2165-2177.

Aral, M. M. M., and Taylor, S. W., 2011, Groundwater quantity and quality management, ASCE Publications.

Archie, G. E., 1942, The electrical resistivity log as an aid in determining some reservoir characteristics: Trans. AIMe, v. 146, no. 99, p. 54-62.

Arnold, J., and Soil, G., 1994, SWAT (Soil and Water Assessment Tool), Grassland, Soil and Water Research Laboratory, USDA, Agricultural Research Service.

Asano, T., and Cotruvo, J. A., 2004, Groundwater recharge with reclaimed municipal wastewater: health and regulatory considerations: Water Research, v. 38, no. 8, p. 1941-1951.

ASTM, 2010, D5447-04 Standard Guide for Application of a Ground-Water Flow Model to a Site-Specific Problem.

Azadpour Keeley, A., Keeley, J. W., Russell, H. H., and Sewell, G. W., 2001, Monitored natural attenuation of contaminants in the subsurface: Processes: Ground Water Monitoring & Remediation, v. 21, p. 97-107.

Barlow, P., DeSimone, L., and Moench, A., 2000, Aquifer response to stream-stage and recharge variations. II. Convolution method and applications: Journal of Hydrology, v. 230, no. 3, p. 211-229.

Barlow, P. M., and Moench, A., 1998, Analytical solutions and computer programs for hydraulic interaction of stream-aquifer systems, US Department of the Interior, US Geological Survey.

Bear, J., 1988, Dynamics of fluids in porous media, Dover publications.

Bear, J., and Verruijt, A., 1987, Modeling groundwater flow and pollution, Springer.

Berryman, J. G., Daily, W. D., Ramirez, A. L., and Roberts, J. J., 2000, Using electrical impedance tomography to map subsurface hydraulic conductivity, Google Patents.

Beven, K., and Binley, A., 1992, Future of distributed models: Model calibration and uncertainty prediction: Hydrological processes, v. 6, no. 3, p. 279-298.

Boel, S., 2008, Hydrogeologische Studie Van De Brownfield-Site Vilvoorde Machelen: Grondwaterstroming En -Transport, MSc Thesis [MSc: Katholieke Universiteit Leuven, 143 p.

Bouwer, H., 2002, Artificial recharge of groundwater: hydrogeology and engineering: Hydrogeology Journal, v. 10, p. 121-142.

Bradley, P. M., and Chapelle, F. H., 1996, Anaerobic mineralization of vinyl chloride in Fe (III)-reducing, aquifer sediments: Environmental Science & Technology, v. 30, p. 2084-2086.

Bredehoeft, J., and Papaopulos, I., 1965, Rates of vertical groundwater movement estimated from the earth's thermal profile: Water resources research, v. 1, no. 2, p. 325-328.

Bren, L., 1993, Riparian zone, stream, and floodplain issues: a review: Journal of Hydrology, v. 150, no. 2, p. 277-299.

Bronders, J., Touchant, K., Van Keer, I., Patyn, J., and Provoost, J., The characterization of contamination and the role of hydrogeology in the risk management of a mega brownfield site, in Proceedings Fifth Congress of the international Association of Hydrologists, Groundwater and ecosystems. Lisbon, Portugal2007.

Bronders, J., Touchat, K., Keer, i. v., Patyn, J., and Provoost, J., A risk managment plan for the redevelopment of a brownfield: an example in Proceedings Proceedings of the 10th International UFZ-Deltares TNO Conference on Soil water Systems 2008, Volume 8, p. 47-54.

Brunke, M., and Gonser, T. O. M., 1997, The ecological significance of exchange processes between rivers and groundwater: Freshwater Biology, v. 37, no. 1, p. 1-33.

Brunner, P., Simmons, C. T., Cook, P. G., and Therrien, R., 2010, Modeling Surface Water Groundwater Interaction with MODFLOW: Some Considerations: Ground Water, v. 48, no. 2, p. 174-180.

Brusseau, M. L., Arnold, R. G., Ela, W., and Field, J., 2001, Overview of innovative remediation approaches for chlorinated solvents: Arizona Department of Environmental Quality, p. 1-63.

Burt, T., 1995, The role of wetlands in runoff generation from headwater catchments, Wiley: Chichester, p. 21-38.

Burt, T., and Pinay, G., 2005, Linking hydrology and biogeochemistry in complex landscapes: Progress in Physical geography, v. 29, no. 3, p. 297-316.

Burt, T., Pinay, G., Matheson, F., Haycock, N., Butturini, A., Clement, J., Danielescu, S., Dowrick, D., Hefting, M., and Hillbricht-Ilkowska, A., 2002, Water table fluctuations in the riparian zone: comparative results from a pan-European experiment: Journal of Hydrology, v. 265, no. 1, p. 129-148.

Burt, T. P., Pinay, G., Sabater, S., International Association of Hydrological, S., Center for Sustainability of semi-Arid, H., and Riparian, A., 2010, Riparian zone hydrology and biogeochemistry, Wallingford, UK, International Association of Hydrological Sciences.

Cartwright, K., 1970, Groundwater discharge in the Illinois Basin as suggested by temperature anomalies: Water resources research, v. 6, no. 3, p. 912-918.

Celia, M. A., Russell, T. F., Herrera, I., and Ewing, R. E., 1990, An Eulerian-Lagrangian localized adjoint method for the advection-diffusion equation: Advances in Water Resources, v. 13, no. 4, p. 187-206.

Cey, E. E., Rudolph, D. L., Aravena, R., and Parkin, G., 1999, Role of the riparian zone in controlling the distribution and fate of agricultural nitrogen near a small stream in southern Ontario: Journal of Contaminant Hydrology, v. 37, p. 45-67.

Chapelle, F., 2001, Ground-water microbiology and geochemistry, Wiley. com.

Chapman, S. W., Parker, B. L., Cherry, J. A., Aravena, R., and Hunkeler, D., 2007, Groundwater–surface water interaction and its role on TCE groundwater plume attenuation: Journal of Contaminant Hydrology, v. 91, no. 3–4, p. 203-232.

Chiang, W. H., and Kinzelbach, W., 1998, Processing Modflow: A simulation program for modelling groundwater flow and pollution. User manual.

Chien, C. C., Medina, M. A., Pinder, J. G. F., Reible, D. D., Sleep, B. E., and Zheng, C., 2004, Contaminated Groundwater and Sediment: Modeling for Mangement and Remediation, Lewis Publihers, p. 274.

Chipperfield, A., and Fleming, P., The MATLAB genetic algorithm toolbox, *in* Proceedings Applied Control Techniques Using MATLAB, IEE Colloquium on1995, IET, p. 10/11-10/14.

Clement, T., Gautam, T., Lee, K., Truex, M., and Davis, G., 2004a, Modeling of DNAPL-dissolution, rate-limited sorption and biodegradation reactions in groundwater systems: Bioremediation journal, v. 8, p. 47-64.

Clement, T., Sun, Y., Hooker, B., and Petersen, J., 1998, Modeling multispecies reactive transport in ground water: Ground Water Monitoring and Remediation, v. 18, no. 2, p. 79-92.

Clement, T. P., 1997, A modular computer code for simulating reactive multi-species transport in 3-dimensional groundwater systems: Pacific Northwest National Laboratory, v. 11720.

Clement, T. P., 1998, RT3D Reaction Module for Modeling Biodegradation Coupled with NAPL Dissolution Processes: Pacific Northwest National Laboratory, Richland, Washington DC.

Clement, T. P., 2011, Complexities in Hindcasting Models—When Should We Say Enough Is Enough?: Ground Water, v. 49, no. 5, p. 620-629.

Clement, T. P., Johnson, C. D., Sun, Y., Klecka, G. M., and Bartlett, C., 2000, Natural attenuation of chlorinated ethene compounds: model development and field-scale application at the Dover site: Journal of Contaminant Hydrology, v. 42, no. 2-4, p. 113-140.

Clement, T. P., Kim, Y.-C., Gautam, T. R., and Lee, K.-K., 2004b, Experimental and numerical investigation of DNAPL dissolution processes in a laboratory aquifer model: Groundwater Monitoring & Remediation, v. 24, p. 88-96.

Commission, E., 2000, Directive of the European Parliament and of the Council 2000/60/EC establishing a framework for Community action in the field of water policy: European Commission PE-CONS, v. 3639, no. 1, p. 00.

Conant Jr, B., 2004, Delineating and quantifying ground water discharge zones using streambed temperatures: Ground Water, v. 42, no. 2, p. 243-257.

Constantz, J., 1998, Interaction between stream temperature, streamflow, and groundwater exchanges in alpine streams: Water resources research, v. 34, no. 7, p. 1609-1615.

Cox, R. A., and Nishikawa, T., 1991, A New Total Variation Diminishing Scheme for the Solution of Advective-Dominant Solute Transport: Water Resources Research, v. 27, no. 10, p. 2645-2654.

De Marsily, G., Delay, F., Gonçalvès, J., Renard, P., Teles, V., and Violette, S., 2005, Dealing with spatial heterogeneity: Hydrogeology Journal, v. 13, p. 161-183.

De Vries, J. J., and Simmers, I., 2002, Groundwater recharge: an overview of processes and challenges: Hydrogeology Journal, v. 10, no. 1, p. 5-17.

Deb, K., Pratap, A., Agarwal, S., and Meyarivan, T., 2002, A fast and elitist multiobjective genetic algorithm: NSGA-II: Evolutionary Computation, IEEE Transactions on, v. 6, p. 182-197.

Desimone, L. A., Barlow, P. M., and Program, G. S. G.-W. R., 1999, Use of computer programs STLK1 and STWT1 for analysis of stream-aquifer hydraulic interaction, US Department of the Interior, US Geological Survey.

Doherty, J., 2008, Addendum to the PEST Manual: Watermark Numerical Computing, Brisbane, Australia.

Doherty, J., and Johnston, J. M., 2003, Methodologies for calibration and predictive analysis of a watershed model1: JAWRA Journal of the American Water Resources Association, v. 39, no. 2, p. 251-265.

Doherty, R. E., 2000, A History of the Production and Use of Carbon Tetrachloride, Tetrachloroethylene, Trichloroethylene and 1, 1, 1-Trichloroethane in the United States: Part 1--Historical Background; Carbon Tetrachloride and Tetrachloroethylene: Environmental forensics, v. 1, no. 2, p. 69-81.

Dou, W., Omran, K., Grimberg, S. J., Denham, M., and Powers, S. E., 2008, Characterization of DNAPL from the U.S. DOE Savannah River Site: Journal of Contaminant Hydrology, v. 97, no. 1-2, p. 75-86.

Douglas, J., Jim, and Russell, T. F., 1982, Numerical methods for convection-dominated diffusion problems based on combining the method of characteristics with finite element or finite difference procedures: SIAM Journal on Numerical Analysis, v. 19, no. 5, p. 871-885.

Dujardin, J., 2012, Detailed Groundwater Recharge Estimation for Urban Environments: a Remote Sensing Supported Procedure for Groundwater Risk Management [PhD: Vrije Universiteit Brussels, 220 p.

Dujardin, J., Batelaan, O., Canters, F., Boel, S., Anibas, C., and Bronders, J., 2011, Improving surface–subsurface water budgeting using high resolution satellite imagery applied on a brownfield: Science of the Total Environment, v. 409, no. 4, p. 800-809.

Ebrahim, G. Y., Hamonts, K., van Griensven, A., Jonoski, A., Dejonghe, W., and Mynett, A., 2012, Effect of temporal resolution of water level and temperature inputs on numerical simulation of groundwater–surface water flux exchange in a heavily modified urban river: Hydrological processes, p. n/a-n/a.

EPA, 2000, Hazard Summary-Created in April 1992; Revised in January 2000, U.S.Environmental Protection Agency.

Essaid, H. I., Zamora, C. M., McCarthy, K. A., Vogel, J. R., and Wilson, J. T., 2008, Using heat to characterize streambed water flux variability in four stream reaches: Journal of environmental quality, v. 37, no. 3, p. 1010-1023.

Evans, R. O., Gilliam, J. W., and Skaggs, R., 1989, Effects of agricultural water table management on drainage water quality.

-, 1991, Controlled drainage management guidelines for improving drainage water quality: AG-North Carolina Agricultural Extension Service, North Carolina State University.

Evans, R. O., Wayne Skaggs, R., and Wendell Gilliam, J., 1995, Controlled versus conventional drainage effects on water quality: Journal of Irrigation and Drainage Engineering, v. 121, no. 4, p. 271-276.

Ferguson, G., and Woodbury, A. D., 2005, The effects of climatic variability on estimates of recharge from temperature profiles: Ground Water, v. 43, no. 6, p. 837-842.

Ferziger, J. H., and Perić, M., 1996, Computational methods for fluid dynamics, Springer Berlin.

Fleckenstein, J. H., Krause, S., Hannah, D. M., and Boano, F., 2010, Groundwater-surface water interactions: new methods and models to improve understanding of processes and dynamics: Advances in Water Resources, v. 33, p. 1291-1295.

Fleckenstein, J. H., Niswonger, R. G., and Fogg, G. E., 2006, River-aquifer interactions, geologic heterogeneity, and low-flow management: Ground Water, v. 44, p. 837-852.

Fohrer, N., and Schmalz, B., 2012, The UNESCO ecohydrology demonstration site Kielstau catchment- sustainable water resources management and education in rural areas: Hydrologie und Wasserbewirtschaftung/Hydrology and Water Resources Management-Germany, v. 56.

Freeze, R. A., and Cherry, J. A., 1977, Groundwater, Prentice-Hall.

Gelhar, L. W., Welty, C., and Rehfeldt, K. R., 1992, A critical review of data on field-scale dispersion in aquifers: Water resources research, v. 28, no. 7, p. 1954-1974.

Geller, J. T., and Hunt, J. R., 1993, Mass transfer from nonaqueous phase organic liquids in water-saturated porous media: Water resources research, v. 29, p. 833-845.

Gen, M., and Cheng, R., 2000, Genetic algorithms and engineering optimization, John Wiley & Sons.

Gorelick, S. M., 1983, A review of distributed parameter groundwater management modeling methods: Water resources research, v. 19, no. 2, p. 305-319.

Gorelick, S. M., Allan Freeze, R., Donohue, D., and Keely, J. o. F., 1993, Groundwater Contamination : Optimal Capture and Containment.

Granholm, J. m., and Chester, S. F., 2003, Preventing Groundwater Contmination: Michigan Department of Environmetal Quality Environmental Science and Services Division: Fact sheet.

Green, J. C., 2006, Effect of macrophyte spatial variability on channel resistance: Advances in Water Resources, v. 29, p. 426-438.

Greenwald, R. M., 1998, Documentation and user's guide: MODMAN, an optimization module for MODFLOW, version 4.0,GeoTrans, HSI.

Gunduz, O., 2006, Surface/subsurface interactions: coupling mechanisms and numerical solution procedures, Groundwater and Ecosystems, Springer, p. 121-130.

Gunduz, O., and Aral, M. M., 2005, River networks and groundwater flow: a simultaneous solution of a coupled system: Journal of Hydrology, v. 301, no. 1, p. 216-234.

Haest, P. J., Springael, D., Seuntjens, P., and Smolders, E., 2012, Self-inhibition can limit biologically enhanced TCE dissolution from a TCE DNAPL: Chemosphere.

Hall, C. W., and Johnson, J. A., 1992, Limiting factors in ground water remediation: Journal of hazardous materials, v. 32, no. 2, p. 215-223.

Hall, F. R., and Moench, A. F., 1972, Application of the convolution equation to stream-aquifer relationships: Water resources research, v. 8, no. 2, p. 487-493.

Hamonts, K., 2009, Structure and pollutant-degrading activity of the microbial community in eutrophic river sediments impacted by discharging chlorinated aliphatic hydrocarbon-polluted groundwater, PhD Thesis [PhD: Katholieke Universiteit Leuven, 209 p.

Hamonts, K., Kuhn, T., Maesen, M., Bronders, J., Lookman, R., Kalka, H., Diels, L., Meckenstock, R. U., Springael, D., and Dejonghe, W., 2009, Factors determining the attenuation of chlorinated aliphatic hydrocarbons in eutrophic river sediment impacted by discharging polluted groundwater: Environmental science & technology, v. 43, no. 14, p. 5270-5275.

Hamonts, K., Kuhn, T., Vos, J., Maesen, M., Kalka, H., Smidt, H., Springael, D., Meckenstock, R. U., and Dejonghe, W., 2012, Temporal variations in natural attenuation of chlorinated aliphatic hydrocarbons in eutrophic river sediments impacted by a contaminated groundwater plume: Water Research, v. 46, no. 6, p. 1873-1888.

Hancock, P. J., 2002, Human impacts on the stream–groundwater exchange zone: Environmental Management, v. 29, p. 763-781.

Hannula, S. R., and Poeter, E., 1995, Temporal and Spatial varations of hydraulic conductivity in a stream bed in Golden Colorado, in Institute, C. W. R. R., ed., Volume Open File Report No.9.

Harbaugh, A. W., Banta, E. R., Hill, M. C., and McDonald, M. G., 2000, MODFLOW-2000, the US Geological Survey modular ground-water model: User guide to modularization concepts and the ground-water flow process, US Geological Survey Reston.

Harvey, J. W., and Bencala, K. E., 1993, The effect of streambed topography on surface-subsurface water exchange in mountain catchments: Water Resources Research, v. 29, p. 89-98.

Hayashi, M., and Rosenberry, D. O., 2002, Effects of Ground Water Exchange on the Hydrology and Ecology of Surface Water: Ground Water, v. 40, no. 3, p. 309-316.

Healy, R., 1990, Simulation of solute transport in variably saturated porous media with supplemental information on modifications to the US Geological Survey's computer program VS2D, Dept. of the Interior, US Geological Survey.

Healy, R., and Ronan, A. D., 1996, Documentation of computer program VS2DH for simulation of energy transport in variably saturated porous media: Modification of the US Geological Survey's computer program VS2DT, US Geological Survey.

Healy, R., and Russell, T., 1993, A finite-volume Eulerian-Lagrangian Localized Adjoint Method for solution of the advection-dispersion equation: Water Resources Research, v. 29, no. 7, p. 2399-2413.

Healy, R. W., 2008, Simulating water, solute, and heat transport in the subsurface with the VS2DI software package: Vadose Zone Journal, v. 7, no. 2, p. 632.

Heyse, E., Augustijn, D., Rao, P. S. C., and Delfino, J. J., 2002, Nonaqueous phase liquid dissolution and soil organic matter sorption in porous media: Review of system similarities: Critical reviews in environmental science and technology, v. 32, p. 337-397.

Hill, A. R., 1996, Nitrate removal in stream riparian zones: Journal of environmental quality, v. 25, no. 4, p. 743-755.

Hinkelmann, R., 2005, Efficient numerical methods and information-processing techniques for modeling hydro-and environmental systems, Springer.

Hou, J., Simons, F., and Hinkelmann, R., 2012, Improved total variation diminishing schemes for advection simulation on arbitrary grids: International Journal for Numerical Methods in Fluids, v. 70, no. 3, p. 359-382.

Hunt, B., 1990, An approximation for the bank storage effect: Water resources research, v. 26, no. 11, p. 2769-2775.

Hunt, J. R., Sitar, N., and Udell, K. S., 1988, Nonaqueous phase liquid transport and cleanup: 1. Analysis of mechanisms: Water Resources Research, v. 24, p. 1247-1258.

Imhoff, P. T., Jaffé, P. R., and Pinder, G. F., 1994, An experimental study of complete dissolution of a nonaqueous phase liquid in saturated porous media: Water Resources Research, v. 30, p. 307-320.

Interstate Technology , Regulatory Cooperation Work Group, and DNAPLs/Chemical Oxidation Work Team, 2000, Dense Non-Aqueous Phase Liquids (DNAPLs):Review of Emerging Characterization and Remediation Technologies.

Jacobs, T., and Gilliam, J., 1985, Riparian losses of nitrate from agricultural drainage waters: Journal of environmental quality, v. 14, no. 4, p. 472-478.

Jones Jr, J. B., and Holmes, R. M., 1996, Surface-subsurface interactions in stream ecosystems: Trends in Ecology & Evolution, v. 11, p. 239-242.

Jonoski, A., Zhou, Y., Nonner, J., and Meijer, S., 1997, Model-aided design and optimization of artificial recharge-pumping systems: Hydrological sciences journal, v. 42, no. 6, p. 937-953.

Kalbus, E., Reinstorf, F., and Schirmer, M., 2006, Measuring methods for groundwater? surface water interactions: a review: Hydrology and Earth System Sciences Discussions, v. 10, p. 873-887.

Kamon, M., Endo, K., Kawabata, J., Inui, T., and Katsumi, T., 2004, Two-dimensional DNAPL migration affected by groundwater flow in unconfined aquifer: Journal of Hazardous Materials, v. 110, no. 1-3, p. 1-12.

Keery, J., Binley, A., Crook, N., and Smith, J. W. N., 2007, Temporal and spatial variability of groundwater-surface water fluxes: Development and application of an analytical method using temperature time series: Journal of Hydrology, v. 336, p. 1-16.

Kendall, C., and MacDonnell, J. J., 1998, Isotope tracers in catchment hydrology, Access Online via Elsevier.

Kennedy, C. A., and Lennox, W. C., 1997, A pore-scale investigation of mass transport from dissolving DNAPL droplets: Journal of contaminant hydrology, v. 24, p. 221-246.

Kipp, K., 1987, HST 3D: A computer code for simulation of heat and solute transport in three dimensional ground-water flow.

Kirschner, E. M., 1994, Environment, health concerns force shift in use of organic solvents: Chemical and Engineering News;(United States), v. 72, no. 25.

Konikow, L. F., 2011, The Secret to Successful Solute-Transport Modeling: Ground Water, v. 49, no. 2, p. 144-159.

Konikow, L. F., and Mercer, J. W., 1988, Groundwater flow and transport modeling: Journal of Hydrology, v. 100, p. 379-409.

Kresic, N., 2006, Hydrofeology and Groundwater Modelling

Kuo, C.-H., Michel, A. N., and Gray, W. G., 1992, Design of optimal pump-and-treat strategies for contaminated groundwater remediation using the simulated annealing algorithm: Advances in water resources, v. 15, no. 2, p. 95-105.

Kwarteng, A. Y., Dorvlo, A. S., and Vijaya Kumar, G. T., 2009, Analysis of a 27-year rainfall data (1977–2003) in the Sultanate of Oman: International Journal of Climatology, v. 29, no. 4, p. 605-617.

Lapham, W. W., 1989, Use of temperature profiles beneath streams to determine rates of vertical ground-water flow and vertical hydraulic conductivity: Available from Books and Open Files Report Section USGS Box 25425, Denver, CO 80225. USGS Water-Supply Paper 2337, 1989. 35 p, 32 fig, 6 tab, 25 ref.

Lappala, E. G., Healy, R. W., and Weeks, E. P., 1987, Documentation of Computer Program VS 2 D to Solve the Equations of Fluid Flow in Variably Saturated Porous Media: Available from Books and Open File Report Section, USGS Box 25425, Denver, CO 80225. USGS Water Resources Investigations Report 83-4099, 1987. 184 p, 27 fig, 15 tab, 51 ref.

LeVeque, R. J., 2002, Finite volume methods for hyperbolic problems, Cambridge university press.

Ling, M., and Rifai, H. S., 2007, Modeling natural attenuation with source control at a chlorinated solvents dry cleaner site: Ground Water Monitoring & Remediation, v. 27, no. 1, p. 108-121.

Littlewood, I. G., and Croke, B. F. W., 2008, Data time-step dependency of conceptual rainfall—streamflow model parameters: an empirical study with implications for regionalisation: Hydrological Sciences Journal, v. 53, p. 685-695.

Littlewood, I. G., Croke, B. F. W., and Young, P. C., 2011, Discussion of "Effects of temporal resolution on hydrological model parameters and its impact on prediction of river discharge": Hydrological Sciences Journal–Journal des Sciences Hydrologiques, v. 56, p. 521-524.

Loke, M., 2000, Electrical imaging surveys for environmental and engineering studies: A practical guide to, v. 2.

-, 2003, RES2DINV v. 3.53. Rapid 2-D resistivity and IP inversion using the least square method: Manual: Penang (Malaysia): Geotomo Software.

Lowrance, R., Leonard, R., and Sheridan, J., 1985, Managing riparian ecosystems to control nonpoint pollution: Journal of Soil and Water Conservation, v. 40, no. 1, p. 87-91.

Lowry, C. S., Walker, J. F., Hunt, R. J., and Anderson, M. P., 2007, Identifying spatial/variability of groundwater discharge in a wetland stream using a distributed temperature sensor: Water resources research, v. 43, p. 10408.

LVA, 2008, Land survey office Schleswig-Holstein 2008, ATKIS©-DEM2 - grid size 1m.

Mackay, D. M., and Cherry, J. A., 1989, Groundwater contamination: Pump-and-treat remediation: Environmental Science & Technology, v. 23, p. 630-636.

Mackay, D. M., Roberts, P. V., and Cherry, J. A., 1985, Transport of organic contaminants in groundwater: Environmental science & technology, v. 19, no. 5, p. 384-392.

Maliva, R. G., and Missimer, T. M., 2010, Aquifer storage and recovery and managed aquifer recharge using wells: Planning, hydrogeology, design, and operation, Schlumberger.

Maskey, S., Jonoski, A., and Solomatine, D. P., 2002, Groundwater remediation strategy using global optimization algorithms: Journal of Water Resources Planning and Management, v. 128, no. 6, p. 431-440.

Mays, L., 2013, Groundwater Resources Sustainability: Past, Present, and Future: Water Resources Management, v. 27, p. 4409-4424.

McCarty, P. L., Biotic and abiotic transformations of chlorinated solvents in ground water1996, p. 5-9.

McCarty, P. L., and Criddle, C. S., 2012, Chemical and biological processes: The need for mixing, Delivery and Mixing in the Subsurface, Springer, p. 7-52.

McKinney, D. C., and Lin, M. D., 1994, Genetic algorithm solution of groundwater management models: Water resources research, v. 30, no. 6, p. 1897-1906.

Mercer, J. W., and Cohen, R. M., 1990, A review of immiscible fluids in the subsurface: Properties, models, characterization and remediation: Journal of Contaminant Hydrology, v. 6, p. 107-163.

Mercer, J. W., and Faust, C. R., 1980, Ground -Water Modeling: An Overviewa: Ground Water, v. 18, no. 2, p. 108-115.

Miller, C. T., Poirier-McNeil, M. M., and Mayer, A. S., 1990, Dissolution of trapped nonaqueous phase liquids: Mass transfer characteristics: Water Resources Research, v. 26, p. 2783-2796.

Moench, A., and Barlow, P., 2000, Aquifer response to stream-stage and recharge variations. I. Analytical step-response functions: Journal of Hydrology, v. 230, no. 3, p. 192-210.

Morrison, R. D., Murphy, B. L., and Doherty, R. E., 2010, 12 Chlorinated Solvents: Environmental Forensics: Contaminant Specific Guide, p. 259.

National Research Council, 1990, GroundWater Models: Scientific and Regulatory Applications, National Academy Press, Washington,D.C.

-, 2000, Natural Attenuation for Groundwater Remediation, National Academies Press.

Neuman, S. P., 1981, A Eulerian-Lagrangian numerical scheme for the dispersion-convection equation using conjugate space-time grids: Journal of computational physics, v. 41, no. 2, p. 270-294.

-, 1984, Adaptive Eulerian–Lagrangian finite element method for advection–dispersion: International journal for numerical methods in engineering, v. 20, no. 2, p. 321-337.

Norris, R. D., and Matthews, J. E., 1994, Handbook of bioremediation, CRC press.

Okeson, S., Iiianagasekare, T. H., Slag, D. C., and Ewing, J. E., Modelling of Dissolution Transport of Nonaqueous Phase Liquid Waste in Heterogeneous

Aquifers, *in* Proceedings Proceedings of the 10th Annual Conference on Hazardous Waste Research, Kansas State University, Manhattan, Kansas, 1995.

Osmond, D. L., Gilliam, J. W., and Evans, R. O., 2002, Riparian buffers and controlled drainage to reduce agricultural nonpoint source pollution, NC Agricultural Research Service, NC State University.

Page, G. W., 1987, Planning for groundwater protection, Access Online via Elsevier.

Panday, Sorab, Langevin, C.D., N., R.G., I., Motomu, and Hughes, J. D., 2013, MODFLOW–USG version 1: An unstructured grid version of MODFLOW for simulating groundwater flow and tightly coupled processes using a control volume finite-difference formulation: U.S. Geological Survey Techniques and Methods, book 6, chap. A45, 66 p.

Pankow, J. F., and Cherry, J. A., 1996, Dense chlorinated solvents and other DNAPLs in groundwater: History, behavior, and remediation.

Papadoulos, S. S., 1980, Evaluation of Alternative Groundwater development schemes for the wadi Samail Aquifer, The Public Autority for Water Resources, Sultanate of Oman, Tetra Tech International

Pereira, L. S., Cordery, I., and Iacovides, I., 2002, Coping with water scarcity. Technical Documents in Hydrology, no. 58, Unesco, Paris.

Peterjohn, W. T., and Correll, D. L., 1984, Nutrient dynamics in an agricultural watershed: observations on the role of a riparian forest: Ecology, v. 65, no. 5, p. 1466-1475.

Pollock, D. W., 1994, User's Guide for MODPATH/MODPATH-PLOT, Version 3: A Particle Tracking Post-processing Package for MODFLOW, the US: Geological Survey Finite-difference Ground-water Flow Model, US Department of Interior.

-, 2012, User guide for MODPATH version 6: a particle tracking model for MODFLOW: US Geological Survey Techniques and Methods, p. 6-A41.

Powers, S. E., Abriola, L. M., and Weber, W. J., 1994, An experimental investigation of nonaqueous phase liquid dissolution in saturated subsurface systems: Transient mass transfer rates: Water Resources Research, v. 30, p. 321-332.

Powers, S. E., Loureiro, C. O., Abriola, L. M., and Weber, W. J., 1991, Theoretical study of the significance of nonequilibrium dissolution of nonaqueous phase liquids in subsurface systems: Water Resources Research, v. 27, p. 463-477.

Prudic, D. E., 1989, Documentation of a computer program to simulate stream-aquifer relations using a modular, finite-difference, ground-water flow model: Available from Books and Open Files Report Section USGS Box 25425, Denver, CO 80225. USGS Open-File Report 88-729, 1989. 113 p, 16 fig, 2 tab, 6 ref, 3 append.

Puckett, L. J., and Hughes, W. B., 2005, Transport and fate of nitrate and pesticides: Journal of environmental quality, v. 34, no. 6, p. 2278-2292.

Putti, M., Yeh, W. W. G., and Mulder, W. A., 1990, A Triangular Finite Volume Approach With High-Resolution Upwind Terms for the Solution of Groundwater Transport Equations: Water Resources Research, v. 26, no. 12, p. 2865-2880.

Pyne, R. D. G., 1995, Groundwater recharge and wells: a guide to aquifer storage recovery, CRC.

Qin, X., and Huang, G., 2009, Characterizing Uncertainties Associated with Contaminant Transport Modeling through a Coupled Fuzzy-Stochastic Approach: Water, Air, & Soil Pollution, v. 197, no. 1, p. 331-348.

Rao, S., Bhallamudi, S. M., Thandaveswara, B., and Mishra, G., 2004, Conjunctive use of surface and groundwater for coastal and deltaic systems: Journal of Water Resources Planning and Management, v. 130, no. 3, p. 255-267.

Rassam, D., Jolly, I., and Pickett, T., 2011, Guidelines for modelling groundwater-surface water interactions in eWater Source: towards best practice model application, eWater Cooperative Research Centre, Canberra, Australia: Interim version, v. 1.

Rassam, D., Pagendam, D., and Hunter, H., 2008, Conceptualisation and application of models for groundwater–surface water interactions and nitrate attenuation potential in riparian zones: Environmental Modelling & Software, v. 23, no. 7, p. 859-875.

Rehg, K. J., Packman, A. I., and Ren, J., 2005, Effects of suspended sediment characteristics and bed sediment transport on streambed clogging: Hydrological Processes, v. 19, p. 413-427.

Reichard, E. G., 1995, Groundwater–surface water management with stochastic surface water supplies: A simulation optimization approach: Water resources research, v. 31, no. 11, p. 2845-2865.

Reichard, E. G., and Johnson, T. A., 2005, Assessment of regional management strategies for controlling seawater intrusion: Journal of Water Resources Planning and Management, v. 131, no. 4, p. 280-291.

Reineke, W., Mandt, C., Kaschabek, S. R., and Pieper, D. H., 2002, Chlorinated hydrocarbon metabolism: eLS.

Richards, L. A., 1931, Capillary conduction of liquids through porous mediums: Physics, v. 1, no. 5, p. 318-333.

Ronan, A. D., Prudic, D. E., Thodal, C. E., and Constantz, J., 1998, Field study and simulation of diurnal temperature effects on infiltration and variably saturated flow beneath an ephemeral stream: Water resources research, v. 34, no. 9, p. 2137-2153.

Rosenberry, D. O., and Pitlick, J., 2009, Effects of sediment transport and seepage direction on hydraulic properties at the sediment–water interface of hyporheic settings: Journal of Hydrology, v. 373, no. 3–4, p. 377-391.

Rosenberry, D. O., Sheibley, R. W., Cox, S. E., Simonds, F. W., and Naftz, D. L., 2013, Temporal variability of exchange between groundwater and surface water based on high-frequency direct measurements of seepage at the sediment-water interface: Water Resources Research, v. 49, p. 2975-2986.

Safavi, H. R., Darzi, F., and Mariño, M. A., 2010, Simulation-optimization modeling of conjunctive use of surface water and groundwater: Water Resources Management, v. 24, no. 10, p. 1965-1988.

Sastry, K., Goldberg, D., and Kendall, G., 2005, Genetic algorithms, Search Methodologies, Springer, p. 97-125.

Scanlon, B. R., Healy, R. W., and Cook, P. G., 2002, Choosing appropriate techniques for quantifying groundwater recharge: Hydrogeology Journal, v. 10, no. 1, p. 18-39.

Schaerlaekens, J., Mallants, D., Simunek, J., van Genuchten, M. T., and Feyen, J., 1999, Numerical simulation of transport and sequential biodegradation of chlorinated aliphatic hydrocarbons using CHAIN_ 2 D: Hydrological processes, v. 13, p. 2847-2859.

Schäfer, W., and Therrien, R., 1995, Simulating transport and removal of xylene during remediation of a sandy aquifer: Journal of Contaminant Hydrology, v. 19, p. 205-236.

Schilling, K. E., Li, Z., and Zhang, Y.-K., 2006, Groundwater–surface water interaction in the riparian zone of an incised channel, Walnut Creek, Iowa: Journal of Hydrology, v. 327, no. 1–2, p. 140-150.

Schmalz, B., Springer, P., and Fohrer, N., 2008a, Interactions between near-surfac groundwater and surfce water in a drained riparian wetland: IAHS Publications-Series of Proceedings and Reports, v. 321, p. 21-29.

-, 2009, Variability of water quality in a riparian wetland with interacting shallow groundwater and surface water: Journal of Plant Nutrition and Soil Science, v. 172, no. 6, p. 757-768.

Schmalz, B., Springer, P., Fohrer, N., Abesser, C., Wagener, T., and Nuetzmann, G., Interactions between near-surface groundwater and surface water in a drained riparian wetland2008b, IAHS Press, p. 21-29.

Schmalz, B., Tavares, F., and Fohrer, N., 2008c, Modelling hydrological processes in mesoscale lowland river basins with SWAT—capabilities and challenges: Hydrological sciences journal, v. 53, no. 5, p. 989-1000.

Schmidt, C., 2009, Water and contaminat fluxes at the stream-groundwater -interface [PhD: University of Neuchâtel.

Schmidt, C., Bayer-Raich, M., and Schirmer, M., 2006, Characterization of spatial heterogeneity of groundwater-stream water interactions using multiple depth streambed temperature measurements at the reach scale: Hydrology and Earth System Sciences Discussions, v. 3, p. 1419-1446.

Schmidt, C., Conant Jr, B., Bayer-Raich, M., and Schirmer, M., 2007, Evaluation and field-scale application of an analytical method to quantify groundwater discharge using mapped streambed temperatures: Journal of Hydrology, v. 347, no. 3-4, p. 292-307.

Schrage, L., 1990, Lindo an Optimization Modeling System/Book and Macintosh Disk, Course Technology Press.

Schreuder, W. A., 2009, BeoPEST Programmer's Documentation, Version 1.0.

Sellmeijer, J., Cools, J., Decker, J., and Post, W., 1995, Hydraulic resistance of steel sheet pile joints: Journal of geotechnical engineering, v. 121, no. 2, p. 105-110.

Semprini, L., Kitanidis, P. K., Kampbell, D. H., and Wilson, J. T., 1995a, Anaerobic transformation of chlorinated aliphatic hydrocarbons in a sand aquifer based on spatial chemical distributions: Water Resources Research, v. 31, p. 1051-1062.

Semprini, L., Kitanidis, P. K., Kampbell, D. H., and Wilson, J. T., 1995b, Anaerobic transformation of chlorinated aliphatic hydrocarbons in a sand aquifer based on spatial chemical distributions: Water resources research, v. 31, no. 4, p. 1051-1062.

Shah, N., Nachabe, M., and Ross, M., 2007, Extinction depth and evapotranspiration from ground water under selected land covers: Ground Water, v. 45, p. 329-338.

Shoemaker, C., Willis, M., and Benekos, I., 2003, Modelling and probabilistic risk analysis of enhanced anaerobic bioremediation of chlorinated ethenes: IAHS publication, p. 374-379.

Siegel, P., Mosé, R., Ackerer, P., and Jaffre, J., 1997, SOLUTION OF THE ADVECTION–DIFFUSION EQUATION USING A COMBINATION OF DISCONTINUOUS AND MIXED FINITE ELEMENTS: International Journal for Numerical Methods in Fluids, v. 24, no. 6, p. 595-613.

Simpson, S. C., and Meixner, T., 2012, Modeling effects of floods on streambed hydraulic conductivity and groundwater-surface water interactions: Water resources research, v. 48, no. 2, p. W02515.

Singh, A., Bürger, C., and Cirpka, O., 2013, Optimized Sustainable Groundwater Extraction Management: General Approach and Application to the City of Lucknow, India: Water Resources Management, v. 27, p. 4349-4368.

Skaggs, R., and Gilliam, J., 1981, Effect of drainage system design and operation on nitrate transport: Transactions of the ASAE [American Society of Agricultural Engineers], v. 24.

Soesilo, J. A., and Wilson, S. R., 1997, Site remediation: planning and management, CRC Press.

Solomatine, D., Dibike, Y., and Kukuric, N., 1999, Automatic calibration of groundwater models using global optimization techniques: Hydrological sciences journal, v. 44, no. 6, p. 879-894.

Sophocleous, M., 2002, Interactions between groundwater and surface water: the state of the science: Hydrogeology Journal, v. 10, no. 1, p. 52-67.

Spanoudaki, K., Nanou-Giannarou, A., Paschalinos, Y., Memos, C., and Stamou, A., 2010, Analytical solutions to the stream-aquifer interaction problem: a critical review: Global Nest Journal, v. 12, no. 2, p. 126-139.

Springer, P., 2006, Analyse der Interaktion zwischen Oberflächenwasser und Grundwasser am Beispiel einer Flussniederung im Norddeutschen Tiefland [MSc: Universität zu Kiel, 191 p.

Srinivas, N., and Deb, K., 1994, Muiltiobjective optimization using nondominated sorting in genetic algorithms: Evolutionary computation, v. 2, p. 221-248.

Stallman, R., 1965, Steady one-dimensional fluid flow in a semi-infinite porous medium with sinusoidal surface temperature: Journal of Geophysical Research, v. 70, no. 12, p. 2821-2827.

Stone, K., Hunt, P., Coffey, S., and Matheny, T., 1995, Water quality status of a USDA water quality demonstration project in the Eastern Coastal Plain: Journal of Soil and Water Conservation, v. 50, no. 5, p. 567-571.

Stone, K., Sommers, R., Williams, G., and Hawkins, D., 1992, Water table management in the Eastern Coastal Plain: Journal of Soil and Water Conservation, v. 47, no. 1, p. 47-51.

Stonestrom, D., and Constantz, J., 2003, Heat as a tool for studying the movement of ground water near streams USGS Circular 1260: US Geological Survey, Denver, CO.

Storey, R. G., Howard, K. W. F., and Williams, D. D., 2003, Factors controlling riffle-scale hyporheic exchange flows and their seasonal changes in a gaining stream: A three-dimensional groundwater flow model: Water Resources Research, v. 39, p. 1034.

Strock, J. S., Kleinman, P. J., King, K. W., and Delgado, J. A., 2010, Drainage water management for water quality protection: Journal of Soil and Water Conservation, v. 65, no. 6, p. 131A-136A.

Su, G. W., Jasperse, J., Seymour, D., and Constantz, J., 2004, Estimation of hydraulic conductivity in an alluvial system using temperatures: Ground Water, v. 42, no. 6-7, p. 890-901.

Surampalli, R. Y., 2004, Natural Attenuation of Hazardous Wastes, ASCE Publications.

Suthersan, S. S., 2010, Natural and enhanced remediation systems, CRC Press.

Suzuki, S., 1960, Percolation measurements based on heat flow through soil with special reference to paddy fields: J. Geophys. Res, v. 65, no. 9, p. 2883–2885.

Taniguchi, M., 2006, Anthropogenic effects on subsurface temperature in Bangkok: Climate of the Past Discussions, v. 2, no. 5, p. 831-846.

Taniguchi, M., Shimada, J., Tanaka, T., Kayane, I., Sakura, Y., Shimano, Y., Dapaah-Siakwan, S., and Kawashima, S., 1999, Disturbances of temperature-depth profiles due to surface climate change and subsurface water flow: 1. An effect of linear increase in surface temperature caused by global warming and urbanization in the Tokyo metropolitan area, Japan: Water resources research, v. 35, no. 5, p. 1507-1517.

Taniguchi, M., and Uemura, T., 2005, Effects of urbanization and groundwater flow on the subsurface temperature in Osaka, Japan: Physics of the Earth and Planetary Interiors, v. 152, no. 4, p. 305-313.

Thomas, D., Hunt, P., and Gilliam, J., 1992, Water table management for water quality improvement: Journal of Soil and Water Conservation, v. 47, no. 1, p. 65-70.

Tiemeyer, B., Frings, J., Kahle, P., Köhne, S., and Lennartz, B., 2007, A comprehensive study of nutrient losses, soil properties and groundwater concentrations in a degraded peatland used as an intensive meadow–implications for re-wetting: Journal of Hydrology, v. 345, no. 1, p. 80-101.

Toro, E. F., 1997, Riemann solvers and numerical methods for fluid dynamics: a practical introduction, Springer.

Tóth, J., 1970, A conceptual model of the groundwater regime and the hydrogeologic environment: Journal of Hydrology, v. 10, p. 164-176.

Touchant, K., Bronders, J., Keer, I. V., Patyn, J., P.Seuntjens, D.Wilezek, and R.Smolders, 2006, Brownfield problematiek Vilvoorde-Machelen; Inventarisatie en karakteristatie van de groundwaterverontreiniging, in VITO, ed.

Triska, F. J., Duff, J. H., and Avanzino, R. J., 1993, The role of water exchange between a stream channel and its hyporheic zone in nitrogen cycling at the terrestrial-aquatic interface: Hydrobiologia, v. 251, no. 1, p. 167-184.

U.S.Epa, 1989, In-Situ Aquifer Restoration of Chlorinated Aliphatics by Methanotrophic Bacteria, EPA/600/2-89/033.

-, 2000, Engineerd Approches to in Situ Bioremediation of Chlorinated Solvents: Fundamentals and Field Applications , EPA-542-R-00-008.

Urish, D. W., 1981, Electrical resistivity—hydraulic conductivity relationships in glacial outwash aquifers: Water resources research, v. 17, no. 5, p. 1401-1408.

van Genuchten, M. T., 1980, A closed-form equation for predicting the hydraulic conductivity of unsaturated soils: Soil Sci. Soc. Am. J, v. 44, no. 5, p. 892-898.

Vogel, T. M., Criddle, C. S., and McCarty, P. L., 1987, ES&T critical reviews: transformations of halogenated aliphatic compounds: Environmental Science & Technology, v. 21, p. 722-736.

Voss, C., 1984, SUTRA: A finite-element simulation model for saturated-unsaturated fluid-density-dependent ground-water flow with energy transport or chemically-reactive single-species solute transport: US Geological Survey Water-Resources Investigations Report, p. 84-4369.

Voudouris, K., 2011, Artificial Recharge via Boreholes Using Treated Wastewater: Possibilities and Prospects: Water, v. 3, no. 4, p. 964-975.

Wagner, B. J., 1995, Recent advances in simulation-optimization groundwater management modeling: Reviews of Geophysics, v. 33, no. S2, p. 1021-1028.

Wang, Y., Bin, H., and Takase, K., 2009, Effects of temporal resolution on hydrological model parameters and its impact on prediction of river discharge: Hydrological Sciences Journal, v. 54, p. 886-898.

Wang, Y., He, B., and Takase, K., 2011, Reply to the Discussion of "Effects of temporal resolution on hydrological model parameters and its impact on prediction of river discharge" by Littlewood et al: Hydrological Sciences Journal–Journal des Sciences Hydrologiques, v. 56, p. 525-528.

Warwick, J., and Hill, A. R., 1988, Nitrate depletion in the riparian zone of a small woodland stream: Hydrobiologia, v. 157, p. 231-240.

Weber, W. J., McGinley, P. M., and Katz, L. E., 1991, Sorption phenomena in subsurface systems: Concepts, models and effects on contaminant fate and transport: Water Research, v. 25, no. 5, p. 499-528.

Wetterhall, F., He, Y., Cloke, H., and Pappenberger, F., 2011, Effects of temporal resolution of input precipitation on the performance of hydrological forecasting: Adv. Geosci, v. 29, p. 21-25.

Whiting, P. J., and Pomeranets, M., 1997, A numerical study of bank storage and its contribution to streamflow: Journal of Hydrology, v. 202, no. 1–4, p. 121-136.

Wiedemeier, T., Swanson, M., Moutoux, D., Gordon, E., Wilson, J., Wilson, B., and KAMPBELL, D. H. H., 1998, Technical Protocol for Evaluating Natural Attenuation of Chlorinated Solvents in Groundwater. National Risk Management Research Laboratory, EPA, Cincinnati, Ohio: EPA/600/R-98/128.

Wiedemeier, T. H., Rifai, H. S., Newell, C. J., and Wilson, J. T., 1999, Natural attenuation of fuels and chlorinated solvents in the subsurface, Wiley.

Wiedemeier, T. H., Wilson, J. T., Kampbell, D., Jansen, J., and Haas, P., Technical protocol for evaluating the natural attenuation of chlorinated ethenes in groundwater1996, p. 425-444.

Winter, T. C., Harvey, J., Franke, O., and Alley, W., 1998, Ground water and surface water: A single resource, Circular 1139: US Geological Survey.

Woessner, W. W., 2000, Stream and Fluvial Plain Ground Water Interactions: Rescaling Hydrogeologic Thought: Ground Water, v. 38, no. 3, p. 423-429.

Woltemade, C., 2000, Ability of restored wetlands to reduce nitrogen and phosphorus concentrations in agricultural drainage water: Journal of Soil and Water Conservation, v. 55, no. 3, p. 303-309.

Wood, R. C., Huang, J., and Goltz, M. N., 2006, Modeling Chlorinated Solvent Bioremediation Using Hydrogen Release Compound (HRC): Bioremediation journal, v. 10, no. 3, p. 129-141.

Zaporozec, A., and Miller, J. C., 2000, Groundwater pollution, Paris,France: UNESCO., UNESCO International Hydrological Programme.

Zedler, J. B., 2003, Wetlands at your service: reducing impacts of agriculture at the watershed scale: Frontiers in Ecology and the Environment, v. 1, no. 2, p. 65-72.

Zekri, S., Ahmed, M., Chaieb, R., and Ghaffour, N., 2013, Managed Aquifer Recharge using quaternary treated wastewater in Muscat: An economic perspective: International Journal of Water Resource Development.

Zektser, I. S., and Lorne, E., 2004, Groundwater resources of the world: and their use: IhP Series on groundwater, no. 6.

Zheng, C., 1990, MT3D: A modular three-dimensional transport model for simulation of advection, dispersion and chemical reactions of contaminants in groundwater systems. US EPA report.

Zheng, C., and Bennett, G. D., 1995, Applied contaminant transport modeling: Theory and practice, Van Nostrand Reinhold New York, v. 5.

Zhou, Y., and Li, W., 2011, A review of regional groundwater flow modeling: Geoscience Frontiers, v. 2, no. 2, p. 205-214.

List of Figures

List of Tables

About the Author

Girma Yimer Ebrahim was born in Wolkite, Ethiopia, on 21st of June 1979. He studied Hydraulic Engineering at Arbaminch Water Technology Institute, Ethiopia, and graduated in 2001 with a BSc. degree in Hydraulic Engineering with Distinction. After his graduation he was employed as a design Engineer at the Irrigation Authority of the Southern Region, Ethiopia, where he worked for three years. During his stay in the organization, he was actively involved in reconnaissance studies, feasibility studies, and the design and evaluation of water resource projects. After he gained the required field experience, he moved to Hawassa University and started to work as graduate assistant involved in teaching regular courses: Hydraulics, Surveying, Engineering Mechanics, Hydraulic and Irrigation Structures. He was also a member of the engineering panel of the university and participated in supervision and design of civil works at the University.

In 2006, he joined UNESCO-IHE for his MSc studies. He obtained the MSc degree in Water Science and Engineering, specialization Hydroinformatics with Distinction in 2008. After that he returned back to his home university and worked as a lecturer in Groundwater Hydrology until he returned to UNESCO-IHE for his PhD studies in May 2009. His research was funded by the EU project AQUAREHAB, a research project initiated in May 2009 with the aim to develop innovative rehabilitation technologies for soil, groundwater and surface water.

JOURNAL PAPERS

Ebrahim, G. Y., Jonoski, A., and Van Griensven, A., (**2009**). Hydrological response of a catchment to climate change in the upper Beles river basin, Upper Blue Nile, Ethiopia: Nile Basin Water Engineering Scientific Magazine, v. 2, p. 49-59.

Ebrahim, G. Y., Hamonts, K., van Griensven, A., Jonoski, A., Dejonghe, W., and Mynett, A., (**2012a**). Effect of temporal resolution of water level and temperature inputs on numerical simulation of groundwater–surface water flux exchange in a heavily modified urban river: Hydrological processes, p. n/a-n/a. DOI: 10.1002/hyp.9310 (IF:2.49)

Ebrahim, G. Y., Jonoski, A., Griensven, A., and Di Baldassarre, G., (**2012b**). Downscaling technique uncertainty in assessing hydrological impact of climate change in the Upper Beles River Basin, Ethiopia: Hydrology Research, doi:10.2166/nh.2012.037

Ebrahim G.Y, Jonoski A., Griensven A., Mynett A., Schmalz B., Fohrer N, Bolekhan A. (**2013**). Quantifying Groundwater - Surface Water Interactions in a Drained Riparian Wetland using Numerical and Analytical Modelling Approaches: under review for Journal of Hydrological processes.

Ebrahim, G. Y., Jonoski, A., Al-Maktoumi, A., Mynett, A., and Ahmed, M., (**2014**) A Simulation-Optimization Approach for Evaluating the Feasibility of Managed Aquifer Recharge in the Samail Lower Catchment, Oman (in preparation).

Ebrahim, G. Y., Jonoski, A., van Griensven, A., Bronders, J., Mynett, A., and Dujardin, J., (**2014**) The Reactive Transport and Natural Attenuation of Chlorinated Solvents at Vilvoorde-Machelen site, Belgium (in preparation).

CONFRENCE PAPERS

Ebrahim, G. Y., Jonoski, A., and Van Griensven, A. (**2008**). Hydrological Response of a Catchment to Climate Change, Case Study on Upper Beles Sub-Basin, Upper Blue Nile, Ethiopia, NBCBN Work shop on the Impact of Climate Change on the Water Resources System of the Nile Basin October 27-28, 2008, Cairo, Egypt.

Corzo, G., Jonoski, A., Ebrahim, G. Y, Xuan, Y., and Solomatine, D. (**2009**). Downscaling global climate models using modular models and fuzzy committees, in Proceedings 8th international conference in hydroinformatics, edited by Liong, S.-Y., World Scientific Publishing Company, Concepcion Chile 2009.

Ebrahim, G. Y., Hamonts, K., Jonoski, A., Dejonghe, W., and van Griensven, A. (**2011**). Modeling groundwater-surface water interaction in the Hyporehic Zone using temperature profiles, HydroEco 2011, Hydrology and Ecology: Ecosystems, Groundwater and Surface Water- Pressure and Options, Vienna, Austria, May 2 - 5, 2011.

Ebrahim, G. Y., van Griensven, A., Jonoski, A., Mynett, A., and Hamonts, K. (**2012**a). Modeling the reactive transport of chlorinated solvents in the groundwater towards the zenne river, Remediation Technologies and their integration in Water Management symposium, September 25-26, 2012, Spain, Barcelona.

Ahmed, M., Al-Maktoomi, A., Ebrahim, G. Y., Slim, Z., and Al-Jabri, S. (**2013**). Storing water underground: A response to anticipated water shortage due to climate change; The threat to Land and Water Resources in the 21st century: Mitgation and Resouration, September 4 -7, Le Meridien Chinag Rai Resort, China.